Microstructure and High-Temperature Performance of Advanced Aluminum Alloys

Ghouse

Table of Contents

Chapter 1

Introduction

1.1 Background

In 21st century, with the rapid growth of economy and society, energy-saving strategies have emphasized a lightweight and eco-friendly approach, particularly in the automotive and aerospace industries [1-4]. As a lightweight material, Aluminium alloys are recognized for their low density, high strength-to-weight ratio, good plasticity, and excellent electrical and thermal conductivity [5-8]. These alloys are extensively used in aerospace, automotive, rail transportation and machinery manufacturing industries since 1970s, replacing cast irons and steels, due to increasing demands on lightweighting and fuel economy. The application of Al alloys in defense structures offers a significant weight advantage, reducing overall weight by 25-40% compared to steel, due to their high capacity for absorbing energy. Figure 1.1 illustrates that Al alloys typically comprise 60-80% of the structural weight in commercial aircraft. While widebody aircraft such as the Boeing 787, Airbus A350, and A380 incorporate a substantial amount of composites (50%, 52%, and 22% by weight, respectively) [9], Al alloys continue to be the primary material in smaller aircraft. The 2xxx, 7xxx, and Al-Li series alloys are most frequently utilized in the aviation industry. However, many critical components operate in high-temperature environments where traditional cast Al alloys degrade in performance. For example, engine combustion chamber parts, exposed to 300-400°C, where prolonged exposure to mechanical and thermal fatigue makes cast Al alloys unsuitable for the demands of high-power engines [10-12].

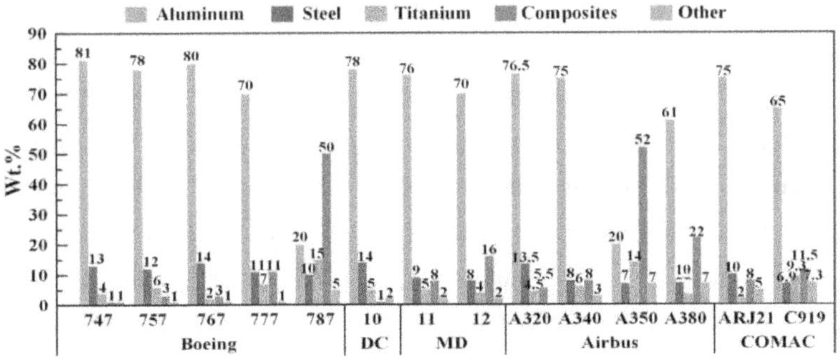

Figure 1.1 The weight percentage of various structural materials used in large commercial aircraft. [9]

In the past decade, heat-resistant cast Al alloys have gained significant attention and seen considerable advancements. The strength enhancement in heat-treatable Al alloys is mainly attributed to the formation of nano-scale precipitates during aging at low temperatures from a supersaturated solid solution. The size, type, orientation, and density of these hardening precipitates have a significant impact on mechanical properties [13-17]. However, it is well understood that as the temperature rises, precipitates coarsen, leading to a steady decline in mechanical properties [12, 18-20]. Figure 1.2 shows the strength reduction in typical heat-treatable 7xxx, 2xxx Al alloys and newly investigated Al-Ni eutectic alloy as temperature increases. A sharp performance drop is observed beyond 200°C, which has hindered the development of lightweight alloys for high-temperature applications. As a result, applications above 250°C pose a significant challenge for heat-treatable Al alloys.

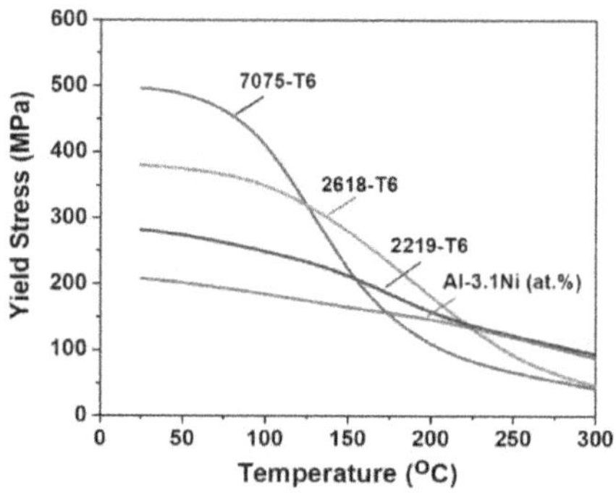

Fig. 1.2 Values of $\sigma_{0.2}$ proof stress of aluminium alloys after thermal exposure [12, 19].

Single solute segregation layers at precipitate/matrix interfaces are not enough to develop heat resistant Al alloys [21-23]. In order to improve the high temperature strength co-precipitation of multiple types of heat-resistant precipitates, intermetallics and segregation structures at precipitate/matrix interfaces are necessary. The significant presence of obstacles within a matrix greatly hinders coarsening phenomena and enhances their strength. Optimizing the composition of principal alloying elements (Zn, Mg, Cu and Ni) plays a key role in the formation of primary strengthening phases to balance the microstructure and mechanical properties. A combination of dispersoid and precipitation hardening concepts through alloying with Mn, Mg, Ni, Cr, Ti, V, Sc and Zr has significantly improved the material tensile strength [24-29]. The alloying elements that produce such coarsening-resistant precipitates in the matrix resulting in a high-volume fraction of the strengthening phase. The $L1_2$ ordered dispersoids at precipitate/matrix interfaces improve the thermal stability of Al alloys [24, 27-30]. The desired microstructure can be achieved if two conditions are met: firstly, the energy between the interfaces of the particles and the matrix must be low, and secondly, the solubility and diffusion rate of the controlling element should be minimal [12, 20, 23-26]. Therefore, these alloys cannot compete with the existing commercial Al alloys in terms of strength. Moreover, the intermetallic and dispersoids have a very strong physio-chemical bond with the Al matrix since they form during solidification.

An alternative approach to refine the microstructures of an alloy is re-melting the surface and sub-surface by LASER or Arc [31-32] due to its high cooling rate (10^4-10^6 K/s). The LASER or Arc acts as the concentrated heat source, penetrating a certain depth in the substrate to melt and solidify the cast microstructure into a different microstructure to enhance mechanical behavior. Like surface remelting, surface alloying by LASER or Arc is another alternative to induce micro elements on the surface to regulate the precipitation behavior, microstructure and mechanical characteristics of the Al alloys [32-33]. Among the different energy sources used for surface modification, the LASER source received the maximum attention from the researchers [31-34]. However, surface modification of Al alloys with LASER is challenging due to the lower melting point of aluminum and the poor absorption of electromagnetic radiation (only 7%) due to the high density of free electrons [32]. Despite these limitations, LASER processing offers improved grain orientation, refinement, and mechanical properties [34-35]. However, the high equipment cost and energy requirements restrict its industrial application [33]. More economical alternatives, such as tungsten inert gas (TIG) or plasma arc, have recently shown promise in refining microstructures and enhancing strength in Al alloys [31, 36-37]. However, with variations in Al alloy composition, the ASR method behaves differently. The Al-Si alloy demonstrated significant grain refinement and precipitation density after ASR, which substantially improves hardness and strength compared to the as-cast condition in an ambient environment [37].

Through an extensive literature review, several inconsistencies have been identified in alloy compositions, aging conditions, operating temperature ranges, manufacturing processing methods and high-temperature characteristics, which constrain the potential of these promising alloys. Furthermore, minimal research has been conducted at 300°C to investigate the phase stabilization of strengthening precipitates by $L1_2$ ordered dispersoids/precipitate/matrix interface and limited research articles investigate the relationship between coarsening behaviour and high-temperature mechanical properties.

In this context, the research idea is to develop a high-performance aluminium alloy that meet the stringent requirements of aerospace applications, providing enhanced strength, thermal stability, and creep resistance under elevated temperatures. The findings will significantly contribute to the design and development of next-generation heat-resistant aluminium alloys for aerospace applications.

In this work, the 7xxx and 2xxx aluminium alloys were processed using a copper-mould gravity die casting route rather than conventional wrought processing. This approach was adopted to systematically isolate and evaluate the intrinsic effects of micro-alloying additions and heat treatment on microstructural stability and elevated-temperature mechanical behaviour, independent of thermomechanical deformation such as rolling or extrusion. While wrought alloys typically undergo DC casting followed by homogenisation and extensive thermo-mechanical processing, the present study focuses on composition and interface-driven strengthening mechanisms applicable to cast or near-net-shape manufacturing routes. Comparisons with conventional wrought alloys such as AA7075 and AA2219 are therefore used as benchmark references to contextualise property trends, rather than as direct processing-route equivalences.

1.2 Scope of the study

- The study focuses on the development of heat-resistant aluminium alloys by investigating the role of micro-alloying elements (Zr, Ti and Sc) in 7xxx (Al-Zn-Mg-Cu) and 2xxx (Al-Cu) series aluminium alloys manufactured by conventional gravity die casting route to enhance their microstructural and elevated temperature mechanical properties.
- The effect of isothermal heat treatments on the microstructure and mechanical properties of micro-alloyed 7xxx and 2xxx series aluminium alloy is investigated.
- The study investigates the effect of micro-alloying elements (Ti, Zr and Sc) on the crack tip damage behaviour of cast 7xxx alloy via in-situ tensile under FESEM.
- The study investigates the impact of arc surface re-melting (ASR) technique on micro-alloyed Al-Cu alloy to understand the influence of surface treatment on grain refinement and high-temperature performance compared to cast materials.
- Optical microscopy, secondary electron (SE) imaging, scanning electron microscopy-energy dispersive spectroscopy (SEM-EDS), scanning electron microscopy-backscattered electron (SEM-BSE), scanning electron microscopy-electron backscatter diffraction (SEM-EBSD), transmission electron microscopy (TEM), transmission electron microscopy-energy dispersive spectroscopy (TEM-EDS), atom probe tomography (APT), differential scanning calorimetry (DSC), and X-ray diffraction (XRD) techniques were used for microstructure analysis.

- Monotonic tensile behaviour at ambient and elevated temperature, bulk-hardness, micro-hardness and tensile creep tests were performed on the alloys.
- Different classical strengthening mechanisms were identified and the role of precipitates, dispersoids and intermetallics in microstructural stability and in strengthening are investigated for micro-alloyed Al-Cu alloy.
- The experimental study was conducted at the following locations,
    - Samples were fabricated at Advanced Casting Research Group, CSIR-CMERI, India.
    - Elevated temperature tensile tests and were conducted at Central Research facility of CSIR-CMERI, India.
    - Surface re-melting treatment was done at Additive manufacturing Research Group, CSIR-CMERI, India.
    - Tensile creep tests were conducted at Advanced Casting Research Group, CSIR-CMERI, India.
    - Defence Metallurgical Research Laboratory (DMRL), India.
    - Metallurgical and Materials Engineering Department at IIT Madras, India.
    - Central Research Facility, IIT Dhanbad, India.
    - RMIT University, Melbourne City campus.
    - The study was conducted from January 2020 to January 2025.

1.3 Aims and Objectives

This PhD work primarily focuses on developing heat-resistant aluminium alloys by investigating the role of micro-alloying constituents in 7xxx series Al-Zn-Mg-Cu and 2xxx series Al-Cu-Mn systems, manufactured using a copper-mould conventional gravity die casting route to systematically investigate composition and interface-controlled strengthening mechanisms relevant to cast and near-net-shape applications. Traditional heat-treatable 7xxx and 2xxx series aluminium alloys experience strength degradation at elevated temperatures due to insufficient stabilization of strengthening precipitates at the precipitate/matrix interfaces. To address this challenge, the research aims to form intermetallics, dispersoids and high-volume fraction of precipitates, particularly through the addition of transition elements, to enhance the high-temperature performance of these conventional alloys. The developed materials will be highly suitable for manufacturing complex components in the aerospace and automotive industries, where both strength and thermal stability are essential. This research addresses the following research questions:

1. What is the impact of precipitate, intermetallic and dispersoids on the microstructure and crack-tip damage mechanisms of base and micro-alloyed Al-6Zn-2.1Mg-2Cu alloy observed during in-situ SEM?

2. What is the role of micro-alloying constituents in 7xxx Series Al-Zn-Mg-Cu alloys:
   (a) How do micro-alloying elements influence the as-cast microstructure and after different iso-thermal heat treatments ?
   (b) What are the effects on ambient and elevated temperature monotonic tensile behaviour ?

3. How does multiple solute segregation at the precipitate/matrix interface contribute to improved thermal stability of Al-Cu-Mn alloy ? A detailed microstructural characterization.

4. What is the impact of micro-alloying constituents in 2xxx Series Al-Cu-Mn alloy:
   (a) How do these constituents influence the as-cast microstructure and respond to iso-thermal heat treatments ?
   (b) How does the tensile behaviour of the alloy at ambient and elevated temperatures compare with existing cast materials designed for high-temperature applications ?
   (c) What are the key factors governing its tensile creep behaviour, including creep rate and failure mechanisms ?

5. What is the impact of Arc Surface Remelting treatment on microalloyed Al-Cu alloy:
   (a) How does the surface treatment affect solute content, nucleation sites, and grain refinement during rapid solidification ?
   (b) Do arc-remelted samples exhibit improved tensile properties at elevated temperatures relative to the as-cast condition? If so, what microstructural mechanisms govern these enhancements ?

1.4 Contribution of the thesis

The research contributes to the development of a high-strength, heat-resistant aluminum alloy, emphasizing the role of multiple solute segregation in stabilizing the microstructure. The study achieves significant strength and thermal stability via incorporating grain boundary (GB) intermetallics, nano-sized precipitates, and $L1_2$-ordered dispersoids. The addition of too much micro-alloying elements in Al-Zn-Mg-Cu alloy forms coarse intermetallics during solidification which agglomerate in the matrix which cause early fracture of the material. The novel micro-alloyed Al-Cu alloy exhibits excellent thermal stability at 300°C, attributed to the presence of multiple solutes, intermetallic phases, and core-shell dispersoids around θ'

precipitates. Arc surface re-melting (ASR) significantly refines the microstructure of Al-Cu alloys and improves their high-temperature properties relative to the as-cast state.

1.5 Thesis Outline

The thesis has been organized into the following nine chapters:

The first chapter introduces the research topic, highlighting the scope and key research questions investigated in this study and the contribution of this work.

The second chapter identifies gaps in the literature on the development of heat-resistant aluminium alloys. A detailed literature review was conducted on alloy design strategies and the role of micro-alloying elements in enhancing high-temperature stability. Key areas include the microstructural and tensile properties of aluminium alloys under aging treatments, and their creep properties at elevated temperatures. The review also focuses on manufacturing process routes for laser- and arc-based surface modification treatments, examining the impact of these methods on the alloys microstructural and mechanical properties.

The third chapter describes methods used for material fabrication, mechanical testing and microstructure quantification in detail.

The fourth chapter focuses on the first research question. It investigates the effect of microalloying elements (Sc, Zr, Ti) on the microstructure and deformation behavior of 7xxx series Al-Zn-Mg-Cu alloys. In-situ tensile testing under FESEM revealed the influence of dispersoid formation on crack initiation, propagation, and fracture mechanisms. The role of optimized microalloying in balancing strength and ductility is emphasized, while excessive additions leading to large intermetallic agglomeration and reduced ductility are also critically analysed.

The fifth chapter address the second research question. It examines the combined effect of controlled microalloying additions and heat treatment strategies (T6, RRA, and HTA) on the microstructure and mechanical performance of Al-Zn-Mg-Cu alloy. This chapter addresses on the precipitation characteristics, phase fractions, and morphologies of the different microalloyed Al-Zn-Mg-Cu alloys. Microstructural changes under elevated temperature testing (200-300 °C) are explored, with emphasis on the coarsening of phases, dynamic softening, and the limitations of thermal stability.

The sixth chapter deals with third research question. It focuses on the effect micro-alloying element on the enhancement of thermal stability in Al-Cu alloy through extensive microstructural characterization and hardness evolution. It examines the role of intermetallics, θ' precipitates, T-$Al_{20}Cu_2Mn_3$ and core-shell $L1_2$-ordered dispersoids on microstructural stability. A detailed microstructural characterization showed coarsening of θ' with respect to aging temperatures.

The seventh chapter deals with the fourth research question. It focuses on the impact of micro-alloying elements and isothermal heat treatments on ambient and elevated temperature tensile and creep properties of the Al-Cu alloy. The role of strengthening mechanisms to the overall yield strength is quantified. Creep mechanisms, stress exponents, and enhanced thermal stability through θ′ precipitate stabilization are thoroughly analyzed. A process-microstructure-property correlation is established for the investigated heat-resistant material.

The eighth chapter deals with the fifth research question. This study evaluates the suitability of micro-alloyed Al-Cu alloy using TIG-assisted arc surface remelting (ASR) to understand the effect of solute content, the role of nucleants and rapid solidification conditions on grain size. The impact of microstructural refinement on ambient and elevated temperatures mechanical properties examines to evaluate the material strength and stability compared to the cast material.

The ninth chapter concludes the research findings of the thesis and recommends future research directions.

References

[1] Zhu, L., Li, N. and Childs, P.R.N., 2018. Light-weighting in aerospace component and system design. Propulsion and Power Research, 7(2), pp.103-119. https://doi.org/10.1016/j.jppr.2018.04.001.

[2] Gayle, F.W. and Goodway, M., 1994. Precipitation hardening in the first aerospace aluminum alloy: the wright flyer crankcase. Science, 266(5187), pp.1015-1017. DOI: 10.1126/science.266.5187.1015.

[3] Wadsworth, J., Nieh, T.G. and Stephens, J.J., 1988. Recent advances in aerospace refractory metal alloys. International materials reviews, 33(1), pp.131-150. doi:10.1179/imr.1988.33.1.131.

[4] Tian, R.Z., 2006. Cast aluminum alloy. Central South University Pres, Changsha.

[5] Liddicoat, P.V., Liao, X.Z., Zhao, Y., Zhu, Y., Murashkin, M.Y., Lavernia, E.J., Valiev, R.Z. and Ringer, S.P., 2010. Nanostructural hierarchy increases the strength of aluminium alloys. Nature communications, 1(1), p.63. https://doi.org/10.1038/ncomms1062.

[6] Hirsch, J. and Al-Samman, T., 2013. Superior light metals by texture engineering: Optimized aluminum and magnesium alloys for automotive applications. Acta Materialia, 61(3), pp.818-843. https://doi.org/10.1016/j.actamat.2012.10.044.

[7] Ma, Y., Addad, A., Ji, G., Zhang, M.X., Lefebvre, W., Chen, Z. and Ji, V., 2020. Atomic-scale investigation of the interface precipitation in a TiB2 nanoparticles reinforced Al-Zn-Mg-Cu matrix composite. Acta Materialia, 185, pp.287-299. https://doi.org/10.1016/j.actamat.2019.11.068.

[8] Wu, S., Soreide, H.S., Chen, B., Bian, J., Yang, C., Li, C., Zhang, P., Cheng, P., Zhang, J., Peng, Y. and Liu, G., 2022. Freezing solute atoms in nanograined aluminum alloys via high-density vacancies. Nature communications, 13(1), p.3495. https://doi.org/10.1038/s41467-022-31222-6.

[9] Li, S.S., Yue, X., Li, Q.Y., Peng, H.L., Dong, B.X., Liu, T.S., Yang, H.Y., Fan, J., Shu, S.L., Qiu, F. and Jiang, Q.C., 2023. Development and applications of aluminum alloys for aerospace industry. Journal of materials research and technology, 27, pp.944-983.

[10] Polmear, I., StJohn, D., Nie, J.F. and Qian, M., 2017. Light alloys: metallurgy of the light metals. Butterworth-Heinemann.

[11] Zhang XM, Liu SD., 2013. Aviation aluminum alloy and itsmaterial processing. Materials China 32, pp.39-55.

[12] Polmear, I.J. and Couper, M.J., 1988. Design and development of an experimental wrought aluminum alloy for use at elevated temperatures. Metallurgical transactions A, 19, pp.1027-1035. https://doi.org/10.1007/BF02628387.

[13] Liu, G., Zhang, G.J., Ding, X.D., Sun, J. and Chen, K.H., 2003. Modeling the strengthening response to aging process of heat-treatable aluminum alloys containing plate/disc-or rod/needle-shaped precipitates. Materials Science and Engineering: A, 344(1-2), pp.113-124. https://doi.org/10.1016/S0921-5093(02)00398-2.

[14] Hornbogen, E., 2001. Hundred years of precipitation hardening. Journal of light metals, 1(2), pp.127-132. https://doi.org/10.1016/S1471-5317(01)00006-2.

[15] Ardell, A.J., 1985. Precipitation hardening. Metallurgical Transactions A, 16, pp.2131-2165. https://doi.org/10.1007/BF02670416.

[16] Kelly, P.M., 1972. The effect of particle shape on dispersion hardening. Scripta Metallurgica, 6(8), pp.647-656. https://doi.org/10.1016/0036-9748(72)90120-2.

[17] Nie, J.F. and Muddle, B.C., 2008. Strengthening of an Al–Cu–Sn alloy by deformation-resistant precipitate plates. Acta Materialia, 56(14), pp.3490-3501. https://doi.org/10.1016/j.actamat.2008.03.028.

[18] Bahl, S., Xiong, L., Allard, L.F., Michi, R.A., Poplawsky, J.D., Chuang, A.C., Singh, D., Watkins, T.R., Shin, D., Haynes, J.A. and Shyam, A., 2021. Aging behaviorand strengthening mechanisms of coarsening resistant metastable θ' precipitates in an Al-Cu

alloy. Materials & Design, 198, p.109378. https://doi.org/10.1016/j.matdes.2020.109378.

[19] Pandey, P., Makineni, S.K., Gault, B. and Chattopadhyay, K., 2019. On the origin of a remarkable increase in the strength and stability of an Al rich Al-Ni eutectic alloy by Zr addition. Acta Materialia, 170, pp.205-217. https://doi.org/10.1016/j.actamat.2019.03.025.

[20] Xue, H., Yang, C., Zhang, P., Wu, S.H., Liu, G. and Sun, J., 2024. Heat-resistant Al alloys: microstructural design and microalloying effect. Journal of Materials Science, 59(22), pp.9749-9767. https://doi.org/10.1007/s10853-023-09295-5.

[21] Nie, J.F., Zhu, Y.M., Liu, J.Z. and Fang, X.Y., 2013. Periodic segregation of solute atoms in fully coherent twin boundaries. Science, 340(6135), pp.957-960. DOI: 10.1126/science.1229369.

[22] Yu, Z., Cantwell, P.R., Gao, Q., Yin, D., Zhang, Y., Zhou, N., Rohrer, G.S., Widom, M., Luo, J. and Harmer, M.P., 2017. Segregation-induced ordered superstructures at general grain boundaries in a nickel-bismuth alloy. Science, 358(6359), pp.97-101. DOI: 10.1126/science.aam8256.

[23] Lu, Q., Wang, J., Li, H., Jin, S., Sha, G., Lu, J., Wang, L., Jin, B., Lan, X., Li, L. and Li, K., 2023. Synergy of multiple precipitate/matrix interface structures for a heat resistant high-strength Al alloy. Nature Communications, 14(1), p.2959. https://doi.org/10.1038/s41467-023-38730-z.

[24] Bansal, U., Singh, M.P., Sinha, S.K., Sahu, D.K., Mondol, S., Makineni, S.K., Paul, A. and Chattopadhyay, K., 2023. Strength and stability through variable micro segregation behaviour of Ta and Zr solutes at intermetallic interfaces in Al-Cu alloys. Acta Materialia, 259, p.119254. https://doi.org/10.1016/j.actamat.2023.119254.

[25] Su, X., Lei, Y., Chen, Y., Qu, H., Qi, Z., Zheng, G., Liu, X., Xiang, H. and Chen, G., 2024. Precipitating thermally reinforcement phase in aluminum alloys for enhanced strength at 400° C. Journal of Materials Science & Technology, 172, pp.71-82. https://doi.org/10.1016/j.jmst.2023.06.038.

[26] Li, R., Takata, N., Suzuki, A. and Kobashi, M., 2022. Design of heat-resistant Al–Mg–Zn–Cu–Ni quinary alloy: controlling intermetallic phases and mechanical performance at elevated temperature. Materials Science and Engineering: A, 857, p.144055. https://doi.org/10.1016/j.msea.2022.144055.

[27] Cassell, A.M., Robson, J.D., Race, C.P., Eggeman, A., Hashimoto, T. and Besel, M., 2019. Dispersoid composition in zirconium containing Al-Zn-Mg-Cu (AA7010)

aluminium alloy. Acta Materialia, 169, pp.135-146. https://doi.org/10.1016/j.actamat.2019.02.047.

[28] Jiang, L., Rouxel, B., Langan, T. and Dorin, T., 2021. Coupled segregation mechanisms of Sc, Zr and Mn at θ′ interfaces enhances the strength and thermal stability of Al-Cu alloys. Acta Materialia, 206, p.116634. https://doi.org/10.1016/j.actamat.2021.116634.

[29] Mondol, S., Kashyap, S., Kumar, S. and Chattopadhyay, K., 2018. Improvement of high temperature strength of 2219 alloy by Sc and Zr addition through a novel three-stage heat treatment route. Materials Science and Engineering: A, 732, pp.157-166. https://doi.org/10.1016/j.msea.2018.07.003.

[30] Poplawsky, J.D., Milligan, B.K., Allard, L.F., Shin, D., Shower, P., Chisholm, M.F. and Shyam, A., 2020. The synergistic role of Mn and Zr/Ti in producing θ′/$L1_2$ co-precipitates in Al-Cu alloys. Acta Materialia, 194, pp.577-586. https://doi.org/10.1016/j.actamat.2020.05.043.

[31] Iwaszko J, Kudła K. Surface remelting treatment of 7075 aluminum alloy—microstructural and technological aspects. Materials Research Express 2020;7:016523. https://doi.org/10.1088/2053-1591/ab625d.

[32] Quazi MM, Fazal MA, Haseeb ASMA, Yusof F, Masjuki HH, Arslan A. Laser-based Surface Modifications of Aluminum and its Alloys. Critical Reviews in Solid State and Materials Sciences 2015;41:106–31. https://doi.org/10.1080/10408436.2015.1076716.

[33] Chi Y, Gu G, Yu H, Chen C. Laser surface alloying on aluminum and its alloys: A review. Optics and Lasers in Engineering 2018;100:23-37. https://doi.org/10.1016/j.optlaseng.2017.07.006.

[34] Ma C, Zhao Q, Liu X, Li Y, Jiang Q. Introducing high-density growth twins in aluminum alloys by laser surface remelting via templated nucleation of grains. Journal of Material Science and Technology 2024. https://doi.org/10.1016/j.jmst.2024.06.036.

[35] Bukhari SMA, Husnain N, Siddiqui FA, Anwar MT, Khosa AA, Imran M, et al. Effect of laser surface remelting on Microstructure, mechanical properties and tribological properties of metals and alloys: A review. Optics & Laser Technology 2023;165:109588. https://doi.org/10.1016/j.optlastec.2023.109588.

[36] Alirezaei M-A, Raiszadeh R, Khayati GR. TIG surface alloying of pure aluminium with manganese. Welding in the World 2023;67:1133-43. https://doi.org/10.1007/s40194-023-01468-9.

[37] Gong Z, Zhang G-D, Zheng F, Wen Z-G, Peng M-C. Performance of different TIG remelting processes in hypereutectic Al-Si cast alloy. Engineering Research Express 2019;1:025027.
https://doi.org/10.1088/2631-8695/ab50f1.

Chapter 2

Literature Review

2.1 Aluminium alloys and its brief history

Today, aluminium is one of the most adaptable, widely used, and affordable metals available. It is astonishing to realize that, about 150 years ago, aluminium was more valuable than gold. Once its unique characteristics were recognized, the metal quickly gained popularity. Aluminium is known for its low density, high tensile strength, malleability, excellent thermal and electrical conductivity, and remarkable resistance to corrosion. In his 1865 novel From the Earth to the Moon, Jules Verne described [1] aluminium as possessing the brightness of silver, the durability of gold, the strength of iron, the melting properties of copper, and the lightness of glass. He noted that it is easy to work with, significantly lighter than iron, and seemed almost tailor-made for ambitious projects. Today, aluminium and its alloys are essential in fields such as construction, transportation, electrical power transmission, packaging, and cookware. Its ability to be recycled efficiently enhances aluminium economic significance and provides substantial environmental benefits by conserving energy [1].

The modern Al industry owes its existence to the simultaneous and independent discoveries made in 1886 by Charles Martin Hall in the United States and Paul Louis Toussaint Héroult in France. They developed the Hall-Héroult process, which revolutionized aluminium production by using electrolysis to extract pure aluminium from alumina (Al_2O_3) dissolved in molten cryolite (Na_3AlF_6). This innovation drastically reduced the cost of aluminium, transforming it from a rare and precious metal into a widely accessible material. To honor the centenary of this breakthrough, France issued a commemorative postage stamp in 1986, as illustrated in Figure 2.1, highlighting the process significance for both industry and society [1]. Shortly after, Austrian chemist Karl Josef Bayer introduced the Bayer process, which efficiently extracts alumina from bauxite ore using caustic soda. The combination of the Bayer and Hall-Héroult processes established the foundation for the large-scale, economical production of aluminium, making it feasible to meet growing industrial demand. The impact of these advancements was quickly evident. Just seven years after the first bulk production of pure aluminium, France erected Anteros, the world's first aluminium statue, in 1893 shown in Figure 2.2 [2]. This statue, representing the angel of Christian charity, showcased the artistic potential of aluminium and symbolized the rapid progress in metallurgy and industrial chemistry. Today, aluminium's versatility and availability stem from these groundbreaking processes, allowing it to be utilized in a vast range of applications and playing a pivotal role in shaping the modern world.

Figure 2.1 In 1886 Paul L. T. Heroult in France and Charles M. Hall in US almost simultaneously devised Aluminium production processes, electrolysis of alumina (Al_2O_3) dissolved in cryolite (Na_3AlF_6)- purification from bauxite [1].

Figure 2.2 Anteros-the world's first aluminium statue, was constructed in 1893 [2].

2.2 Primitive alloys

The creation of bronze, achieved by alloying tin and copper around five thousand years ago, marked a significant milestone in material development. Since then, the practice of combining

base metals has led to the creation of numerous new materials. With the advent of commercial aluminum production over a century ago, aluminum has served as the foundation for a wide range of alloys. The serendipitous discovery of age hardening in copper-containing aluminum alloys by Alfred Wilm in 1906 [3] significantly increased the strength of this lightweight metal. The inherent low density of aluminium alloys subsequently secured their prominence in aerospace engineering and premium automotive manufacturing. Continuous advancements in alloy design have pushed conventional aluminum alloys to their performance boundaries. However, aluminum now faces stiff competition from alternative materials, such as lightweight polymer composites, and its use in high-temperature environments remains limited [4].
Aluminum is alloyed with multiple alloying elements, resulting in major alloy series such as 2XXX (Al-Cu), 3XXX (Al-Mn), 5XXX (Al-Mg), 6XXX (Al-Mg-Si), and 7XXX (Al-Zn-Mg-Cu), each offering distinct combinations of strength and lightweight properties [4]. Despite their advantages, many of these alloys lose strength at elevated temperatures, often due to the coarsening of precipitates or dispersed phases. Enhancing high-temperature performance has therefore remained a central focus in aluminum alloy research for decades. One promising approach involves creating stable, nanoscale dispersions during rapid solidification, which can help maintain strength at higher temperatures.

2.3 Aluminium alloys as heat-resistant structural materials

The lightweight heat-resistant alloys are essential for transportation, where the goal is to achieve a strength-to-weight ratio comparable to that of steel or titanium alloys, often assessed by aiming for similar specific modulus values (E/ρ) in aluminium alloys. On the other hand, thermal stability in structural materials lacks a singular definition but is commonly understood as the capacity to retain properties at elevated temperatures over prolonged periods; however, it encompasses key parameters including temperature, time, loading conditions, and environmental factors. Alternative interpretations, such as resistance to permanent heat-induced property changes, are imprecise, as some recovery often occurs upon cooling, though engineering designs prioritize in-service performance at operating temperatures. Thermal stability is sometimes treated as an intrinsic property characterizing long-term high-temperature exposure, yet this overlooks application-specific nuances, while "dimensional thermal stability" specifically addresses thermal expansion [5]. Thermal stability evaluation varies by context: temperature and time are sufficient for comparing alloys in development, but engineering material selection requires comprehensive tests with load and environmental simulations. The relatively low melting point of Al (660.5°C) restricts the thermal stability of its alloys to lower temperature regimes, in contrast to materials with higher melting points

(Figure 2.3) [6]. Compared with steels and Ti alloys, Al alloys demonstrate distinct property changes when exposed to temperatures ranging from cryogenic levels to above 300°C; these variations are significantly influenced by their composition, microstructure, manufacturing routes, and post-processing.

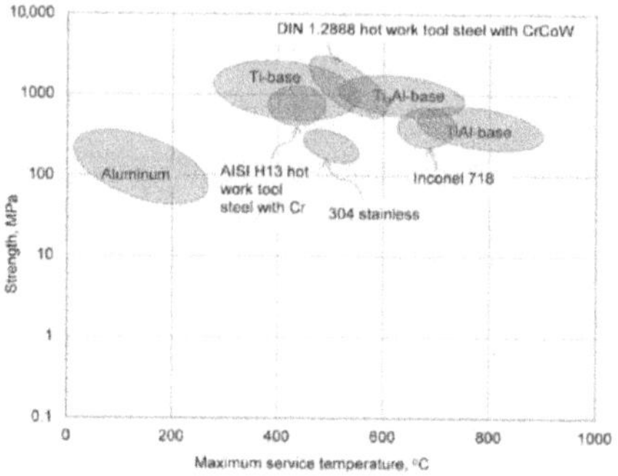

Figure 2.3 Strength versus maximum operating temperature for Al alloys and other structural materials [6].

For instance, in 7075 and 2219 alloys under T6/T651 conditions, exposure above 150°C leads to progressive loss of strength that becomes more pronounced with increasing time and temperature. When the temperature exceeds 200°C, significant softening is observed along with increased ductility, and most of the strength reduction is irreversible due to the coarsening of Mg_2Zn and Al_2Cu precipitates. This coarsening is thermodynamically driven by minimization of interfacial energy. To enhance high-temperature stability, various processing strategies are employed, such as optimizing precipitate distributions or incorporating stabilizing elements. Though Al alloys remain prone to creep which is a time-dependent deformation under sustained loads below yield strength, primarily facilitated by thermally activated vacancy migration. These concepts underscore the multifaceted design challenges in developing heat-resistant Al alloys for aerospace applications.

2.4 High-temperature degradation of alloy strengthening phases

Increasing operating temperatures introduce additional challenges for alloys beyond those seen at room temperature. Achieving this improvement critically depends on understanding the intricate interplay between an alloy chemical composition, processing parameters, microstructural evolution, and its properties under service conditions. Advances in alloy performance over recent decades are largely attributed to deeper insights into these structure-property relationships. Incorporating alloying elements into Al provides multiple beneficial effects, including precipitation hardening, solid solution strengthening, dispersion hardening, grain refinement, modification of metallic and intermetallic phases and suppression of grain growth at high temperatures.

The elastic behaviour of materials provides insight into their mechanical characteristics, including ductility, toughness, hardness, and strength. The Young's modulus can be modified through alloying, and the extent of variation depends on both the concentration of alloying elements and the way they are incorporated. When solute atoms remain in solid solution, the modulus is primarily governed by atomic bonding interactions (Figure 2.4). In contrast, when alloying elements form secondary phases, the effective Young's modulus is influenced by the intrinsic stiffness of those phases and their relative volume fraction within the alloy [7].

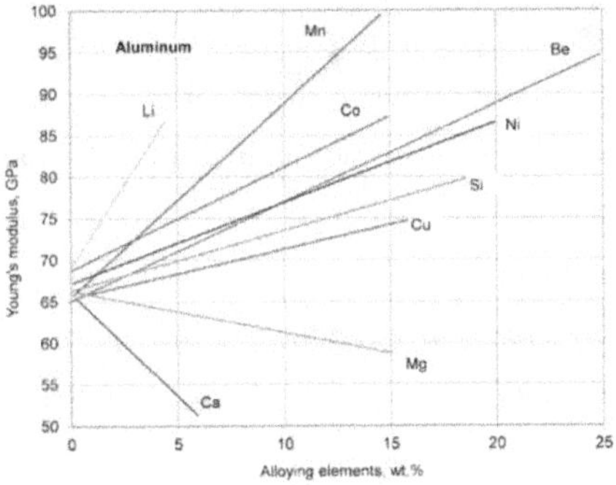

Figure 2.4 Effect of solute additions on the elastic modulus of Al alloys, adapted from the findings reported in [7].

2.5 Strengthening mechanisms in alloys

The thermal stability of an alloy is governed by microstructural changes induced by the diffusion of alloying elements. As the temperature increases, microstructural features-such as grain size, phase distribution, morphology, and the concentration of solute atoms in the matrix begin to evolve. To evaluate the factors influencing thermal stability, it is important to consider the primary strengthening mechanisms: grain boundary strengthening, solid solution strengthening, second-phase precipitation, and work hardening operating in Al alloys (Figure 2.5) [8]. To enhance these mechanisms, researchers have developed various approaches, such as modifying alloy compositions, refining heat treatment processes, and incorporating reinforcing elements. However, when exposed to elevated temperatures, the mechanisms that contribute to strength at ambient conditions may vary in their effectiveness, thereby impacting the overall strength of the alloy.

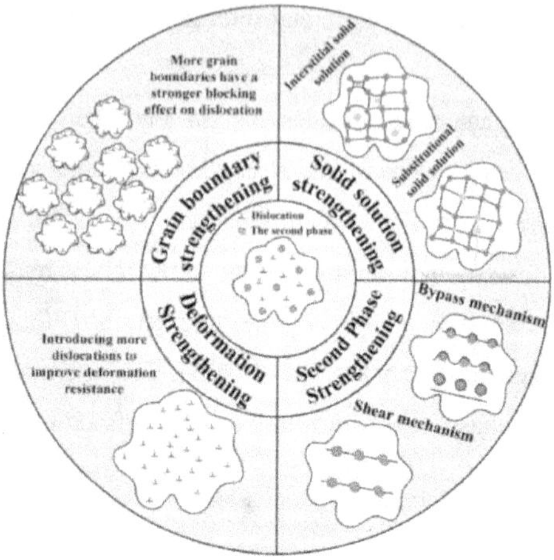

Figure 2.5 Strengthening mechanisms for aluminium alloys in aerospace applications [8].

2.5.1 Solid solution strengthening

Solute atoms enhance alloy strength by introducing lattice distortions in the crystal structure, which in turn impedes the motion of dislocations. These atoms tend to migrate and accumulate near dislocation lines, where they can occupy more energetically favourable positions based on their atomic radii. This segregation reduces the overall strain energy and effectively restricts dislocation movement, which is known as dislocation pinning. The strengthening effect due to solid solution hardening is primarily governed by the strain fields generated around individual solute atoms, which depend on the mismatch in atomic size between solute and solvent atoms. Several factors contribute to solid solution strengthening, including differences in atomic size, elastic modulus, chemical affinity, and configurational arrangement. Among these, the relative size factor (denoted as "ε_a") is defined by the following expression [9, 10]:

$$\varepsilon_a = \frac{1}{a}\frac{da}{dc} \tag{1}$$

where, *a* represents the interatomic spacing of the alloy, and *c* is the concentration of the solute atoms.

The strengthening also depends upon the concentration of solute in the solid solution which is typically represented as [10],

$$\tau_{ss} = A{C_o}^{2/3} \tag{2}$$

where τ_{ss} is increase in yield stress due to solid solution strengthening, *A* the constant and C_o is concentration of the solute in weight percent.

2.5.2 Precipitate strengthening (second phase strengthening)

One of the most efficient strategies to achieve high yield strength in metallic alloys is through dispersion and precipitation hardening. The theoretical upper limit of yield stress can be estimated using the relation: $\sigma_{theoritical} \geq \frac{G_{Al}}{10\pi}$. For aluminium, this value is ~900 MPa. However, in practical, the actual yield strength is considerably lower due to defects, particularly dislocations, which facilitate plastic deformation at much lower stress levels. Strengthening can be significantly improved by introducing a uniform distribution of precipitates or dispersed particles within the matrix. These second-phase particles act as obstacles to dislocation motion, increasing the resistance to deformation. The strengthening efficiency of these particles largely depends on their size, with a critical particle size (d_c) required for effective strengthening, depending on the nature of the particles (Figure 2.6) [11]. These are following:

Ordered coherent particles: $d_C = \frac{G_\alpha b^2}{\gamma_{APB}}$ (3)

Disordered coherent particles: $d_c = \frac{G_\alpha b}{\tau_\alpha - \tau_\beta}$ (4)

For incoherent precipitates with different crystal structures:

$$d_c = \frac{4\pi G_\alpha b}{G_\beta} \quad (5)$$

Here, G is the shear modulus, b is the Burgers vector, γ_{APB} is the anti-phase boundary energy, and $(\tau_\alpha - \tau_\beta)$ represents the difference in shear stress between the α and β phases.

Due to the limited solubility of alloying elements in aluminium, it is essential to analyze the quantitative influence of both the volume fraction and distribution of precipitated or dispersed particles. When precipitates are fine, coherent, and shearable, strengthening arises from the resistance offered by precipitates to dislocation cutting. The increase in yield strength due to precipitate shearing can be expressed as:

$$\Delta\sigma_{shear} = M\alpha Gb\sqrt{\frac{f}{r}} \quad (6)$$

where f is the precipitate volume fraction, r is the average precipitate radius, and α is a strengthening coefficient that depends on precipitate-matrix coherency and modulus mismatch. This mechanism dominates at lower aging temperatures and early aging stages, where precipitates remain coherent and uniformly distributed.

The Orowan mechanism describes the strengthening effect caused by dislocations bypassing non-shearable particles. The stress required for a dislocation to loop around such particles is inversely related to the inter-particle spacing. The increase in yield strength [12] due to the Orowan looping mechanism is given by:

$$\Delta\sigma_{orowan} = M\frac{0{\cdot}4Gb}{\pi\sqrt{(1-\vartheta)}}\frac{ln\left(\frac{2\bar{R}}{b}\right)}{\lambda} \quad (7)$$

Where the spacing and size relationships are: $\bar{R} = \frac{\pi}{4}\langle R\rangle$ (8)

$$\lambda = \left(\sqrt{\frac{2\pi}{3f}} - 2\frac{\pi}{.4}\right)\langle R\rangle \quad (9)$$

Where, M is the Taylor factor, G and ϑ are the shear modulus and Poisson's ratio of the matrix respectively, b is the magnitude of the Burgers vector, $\bar{R}$ is mean planar precipitate radius, $\langle R\rangle$ is the average particle radius, λ is the effective interparticle spacing, and f is the volume fraction of precipitates or dispersoids. These equations provide a reasonable approximation for systems with a distribution of particle sizes [12]. A graphical representation of Orowan stress as a function of precipitate radius for various volume fractions, based on aluminium alloys, is shown in Figure 2.7 and is compared with Ni-based alloy systems.

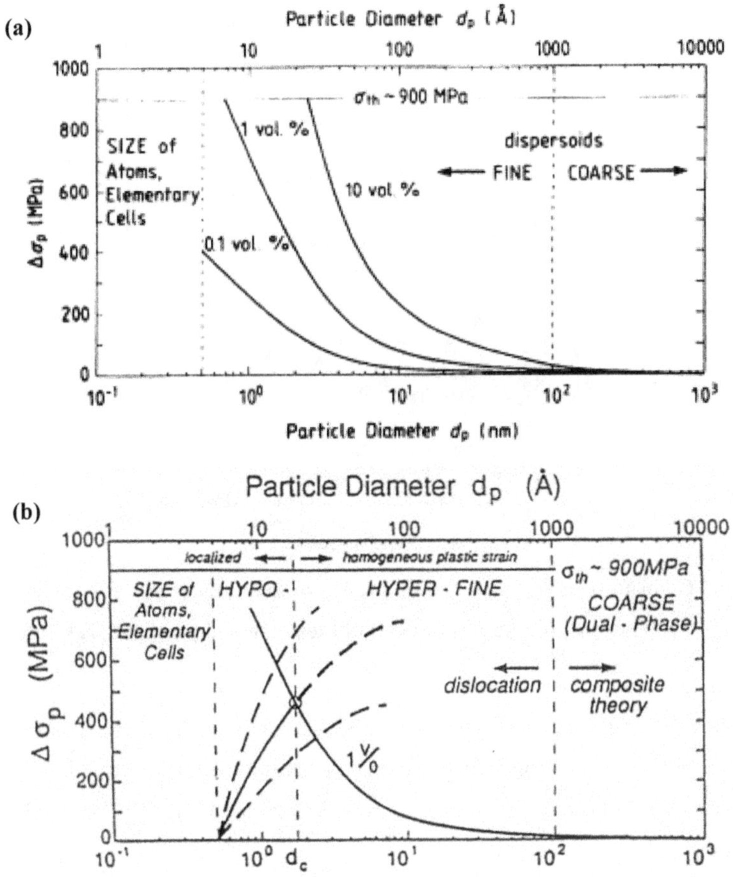

Figure 2.6 (a) Calculated yield strength improvement in Al-alloys by dispersed particles; (b) Hornbogen and Starke [58] illustrated the theoretical relationships between precipitation hardening (σ_p) and both the volume fraction and diameter of the particles. The critical particle diameter (d_c) threshold for transitioning from dispersion shearing in fine microstructures (dislocations penetrate and shear small particles, contributing to strength via coherency strains) to bypassing in coarse microstructures (dislocations detour around larger particles, forming loops as described by dislocation theory) [11].

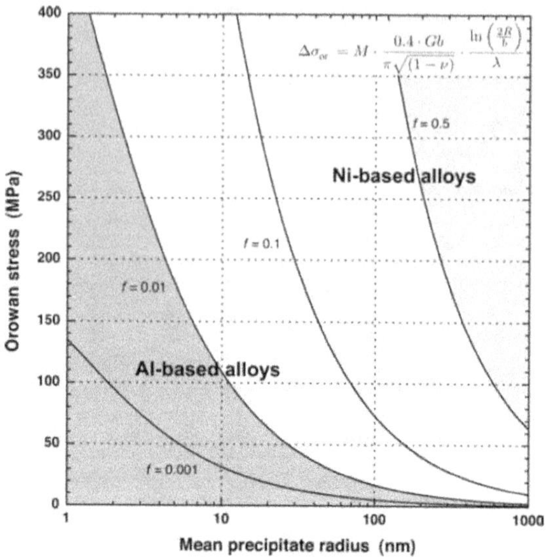

Figure 2.7 Orowan stress is plotted as a function of the average precipitate radius ⟨R⟩ for varying volume fractions (f) of the dispersed particles [12].

2.5.3 Grain boundary strengthening

When plastic deformation is mainly controlled by dislocation glide and diffusion effects are minimal, a reduction in grain size leads to enhanced material strength. Under such circumstances, the yield strength due to grain boundaries (σ_{gb}) is described by the Hall-Petch relationship [5]:

$$\sigma_{gb} = \sigma_0 + kd^{-1/2} \tag{10}$$

Here, σ_0 represents the intrinsic resistance of the crystal lattice to dislocation motion, and k is a material-specific strengthening coefficient, which is smaller for Al alloys than other alloy systems.

An alternative approach to describe grain boundary strengthening in aluminium alloys is given by the Nes-Marthinsen equation [13]:

$$\sigma_{gb} = \alpha G b d^{-1} \tag{11}$$

where, G is the shear modulus, b is the Burgers vector, and d is the average grain size, with the proportionality constant dependent on material characteristics [13]. It is important to note that aluminium alloys are influenced by multiple strengthening mechanisms acting in tandem. As a result, the mechanical properties during service conditions is typically a combined effect of different strengthening contributions [14].

2.5.4 Dislocation strengthening

Dislocation strengthening is a fundamental mechanism that enhances the mechanical properties of cast aluminum alloys by impeding dislocation motion through interactions with other dislocations and microstructural features [15]. According to Taylor's model, the strengthening effect can be quantified by the relationship:

$$\Delta\tau_{DIS} = M\alpha Gb\sqrt{\rho}_{DIS} \tag{12}$$

where $\Delta\tau_{DIS}$ represents the increase in yield strength, α is a material-specific constant (typically 0.2-0.5 for Al alloys), G is the shear modulus (~26 GPa for Al), b is the Burgers vector magnitude (~0.286 nm), and ρ is the dislocation density [16, 17]. In cast Al alloys, dislocations naturally form during solidification and processing, and their density can be further increased through mechanical working. As dislocations multiply and interact, they create a network of obstacles that restrict further dislocation glide, thereby increasing the material's resistance to deformation. This strengthening mechanism is particularly important in cast alloys where other strengthening methods like precipitation hardening may be limited. However, excessive dislocation accumulation can lead to reduced ductility, necessitating careful control of processing parameters to optimize the balance between strength and toughness in cast aluminium components [18].

2.6 Precipitate stability

When heat-treatable Al alloys are exposed to elevated temperatures, the Gibbs-Thomson effect drives the transformation of high-density tiny precipitates into fewer, larger ones, thereby reducing the overall interfacial area [19]. This coarsening process accelerates with increasing temperature, leading to an undesirable decline in material strength [20]. As a result, controlling the coarsening of precipitates is a key challenge in developing materials intended for high-temperature environments. The Lifshitz–Slyozov–Wagner (LSW) model describes the growth of average precipitate size in a dilute binary system follows this equation [21,22]:

$$(r_t)^n - (r_0)^n = \frac{8C_i^m(1-C_i^m)\gamma D V_m}{9RT\left(c_i^P - c_i^m\right)^2} \times (t - t_0) \tag{13}$$

where, r_t and r_0 represent the precipitate radius at times t and t_0, respectively; n denotes the diffusion mechanism index (the volume diffusion type $n = 3$ and the interface diffusion type $n = 2$),); D is the diffusion coefficient of the solute in the Al matrix; γ is the interfacial energy between precipitates and the matrix; T is the absolute temperature; c_i^P and c_i^m are the solute concentrations in the precipitates and matrix, respectively; V_m is the molar volume of the precipitates; and R is the gas constant (8.31 J/mol·K). According to this relationship, four main factors are identified as having a significant influence on precipitate coarsening. First, the size distribution of the precipitates: a more uniform distribution can lower the driving force for coarsening, while a broad distribution with many large and small particles tends to accelerate it. By carefully managing both nucleation and growth, it is possible to achieve a narrower size distribution and thus slow down coarsening. Second, the interfacial energy between the precipitate and the Al matrix is important; fully coherent interfaces have low interfacial energy, which helps stabilize the precipitates at high temperatures. Third, the rate at which elements diffuse through the Al matrix; limiting this diffusion can help suppress coarsening. Fourth, the solubility of the element in the matrix is crucial; lower solubility at service temperatures helps prevent the dissolution and subsequent coarsening of the precipitates.

At temperatures exceeding 350°C, metastable precipitates not only coarsen and convert to stable equilibrium phases, which adversely affects mechanical properties. In Al-Cu alloys, the high density of metastable θ'-Al_2Cu precipitates is primarily responsible for the alloy's strength [23-25]. These plate-like θ'-Al_2Cu precipitates develop on $(100)_{\alpha\text{-Al}}$ planes, featuring broad coherent interfaces and semi-coherent side interfaces (with lattice parameters a = 0.404 nm and c = 0.580 nm) [26, 27]. However, when the temperature exceeds 300°C, these metastable θ'-Al_2Cu precipitates begin to coarsen, their number density drops, and their aspect ratio decreases, eventually leading to their transformation into the stable θ-Al_2Cu phase. This phase transition results in a significant decline in mechanical performance. Therefore, restricting the evolution from metastable to equilibrium phases is essential for extending the high-temperature capabilities of Al alloys.

2.7 Diffusivity of alloying elements in Al

Alloy stability during service critically depends on the diffusivity of alloying elements. For substitutional solutes, diffusion occurs predominantly through the migration of atomic lattice vacancies, often necessitating nonequilibrium vacancy concentrations to drive structural evolution. Aluminum, a trivalent face-centered cubic (fcc) metal, exhibits distinct impurity diffusion characteristics compared to monovalent fcc noble metals. Notably, impurity elements exhibiting low solid solubility in Al typically possess diffusion coefficients significantly lower than that of Al self-diffusion. Figures 2.8 and 2.9 illustrate this comparison for selected alloying elements diffusing in Al [6].

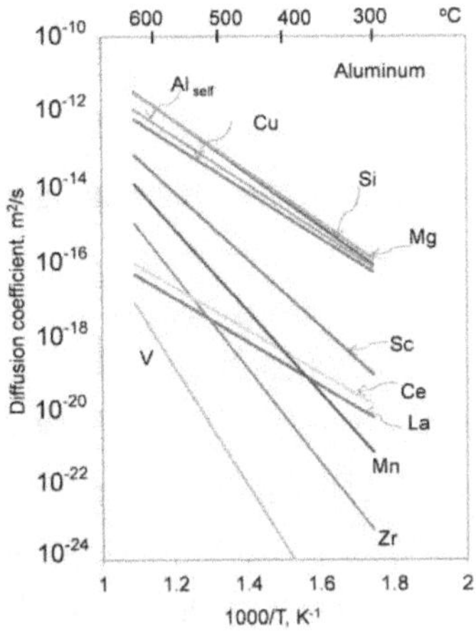

Figure 2.8 Semi-logarithmic representation of diffusivity in Al as a function of reciprocal temperature, comparing Al self-diffusion with alloying elements [6].

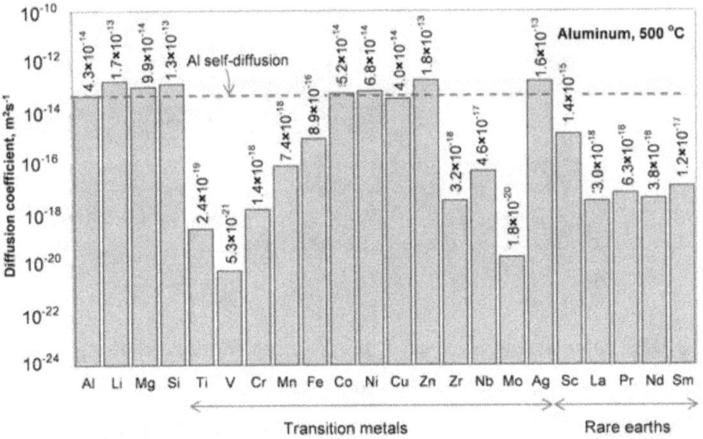

Figure 2.9 Diffusion coefficients at 500 °C for selected alloying elements in Al [6].

Most transition metals exhibit lower diffusivities than Al self-diffusion. Typically, these elements possess high activation enthalpies and pre-exponential factors. For partially filled d-shell 3d elements, elevated activation barriers and anomalously low diffusion coefficients signify strong bonding with Al [28]. Co, Ni, and Cu have diffusivities close to that of Al; Zn and Ag diffuse faster, while Fe, Mn, Cr, Ti, Zr, V, Hf, Mo, and Pd exhibit lower diffusion rates [29, 30].

Rare earth (RE) metals have minimal solid-state solubility in Al due to their large atomic size and trivalent charge. Diffusion data for these elements is limited, with values primarily sourced from a 1960s study [31]. Typically, RE diffusion coefficients in Al are up to four orders of magnitude lower than Al self-diffusion coefficients. Scandium is an exception, its diffusion rate lying midway between that of other rare earths and Al (Figure 2.8 and 2.9).

2.8 Exploring novel strengthening dispersions in lightweight Al Alloys

Recent advances in materials discovery for lightweight Al alloys are increasingly driven by computational modeling, simulation, and validation using extensive databases. High-throughput density functional theory (HT DFT) calculations, such as those available through the Open Quantum Materials Database (OQMD, with over 200,000 entries at http://oqmd.org), have become efficient tools for accelerating the design and identification of new materials. These DFT databases are commonly used to predict crystal structures and guide alloy development. For example, in Mg alloys, computational data mining facilitated by the Materials Genome Initiative predicted the strengthening effect of stable ternary long-period stacking ordered (LPSO) precipitates, as shown in Figure 2.10 [32, 33].

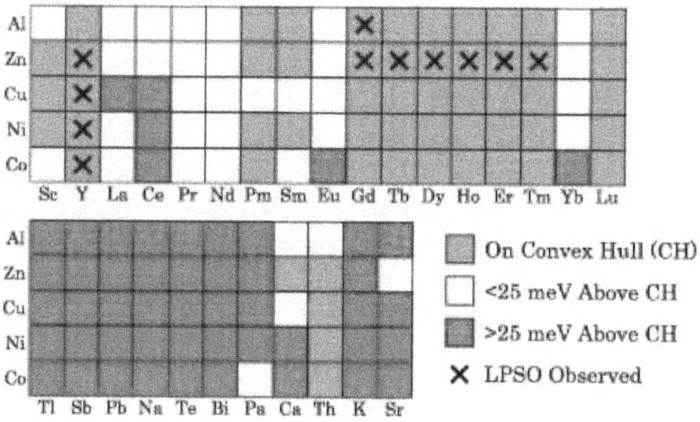

Figure 2.10 DFT calculations were used to assess the stability of Long Period Stacking Ordered (LPSO) phases in Mg-XL-XS ternary alloys. In the corresponding diagram, XL DFT calculations were used to assess the stability of Long Period Stacking Ordered (LPSO) in Mg-XL-XS ternary alloys. XL and XS elements are shown on the vertical and horizontal axes, respectively. The colors coding indicates LPSO stability relative to the convex hull: yellow for systems within 25 meV/atom above the hull, and red for those exceeding 25 meV/atom. Alloys with X^L = RE are displayed at the top, while those with $X^L \neq$ RE are at the bottom. Experimentally confirmed LPSO-forming systems are marked with "×". Blue and possibly yellow squares without "×" highlight predicted alloy systems where LPSO phases are expected but have not yet been experimentally confirmed [32, 33].

2.9 Alloying elements

2.9.1 Thermodynamics of Sc addition

Adding scandium (Sc) to cast Al alloys improves specific strength, and thermal stability [34-37]. The binary phase diagram (Figure 2.11a) of Al-Sc alloy shows a eutectic point at ~0.56 wt.% Sc and a eutectic temperature of ~659°C, which is just below the melting point of pure Al (~660.3 °C) [35, 38, 39]. This results in a narrow solidification range, unlike the broader intervals (~90-130 °C) seen in conventional 2xxx and 7xxx series Al alloys [40-42]. The narrow range, combined with Sc's low diffusion rate in Al, promotes the formation of a supersaturated solid solution of Sc in Al during solidification. Under equilibrium conditions, the solid solubility of Sc in Al is limited to ~0.35-0.38 wt.%, but this can be extended thermodynamically through non-equilibrium cooling [38, 39].

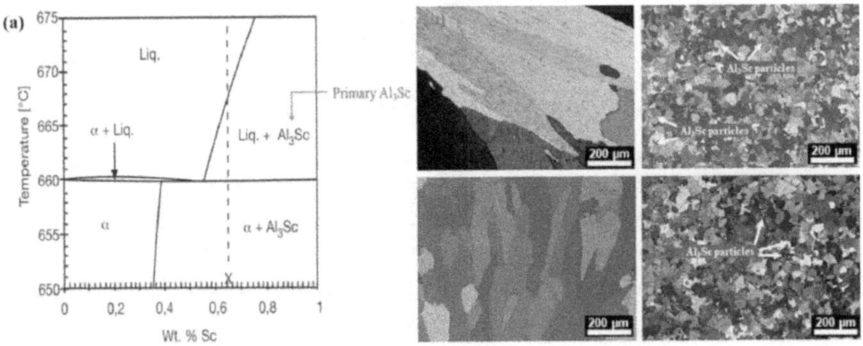

Figure 2.11 (a) Al-Sc equilibrium phase diagram [38]; optical micrographs of as-cast samples (b) pure Al, (c) Al-0.5 wt.% Sc, (d) Al-0.7 wt.% Sc, and (e) Al-1.0 wt.% Sc [43].

In hypereutectic Sc-cast Al alloys (Sc > 0.6 wt.%), primary Al_3Sc or $Al_3(Sc_{1-x}TM_x)$ intermetallics (when alloyed with transition metals like Zr or Ti) form first in the melt, acting as heterogeneous nucleation sites for α-Al grains. The precipitation temperature of these Al_3TM can reach up to ~800 °C, depending on composition. Thermodynamically, the effectiveness stems from the close lattice match between α-Al (FCC, 0.404 nm) and Al_3Sc phases (cubic $L1_2$, ~0.4104 nm), reducing the energy barrier for nucleation and promoting equiaxed, fine-grain structures [40-44]. Figure 2.11(b-e) illustrates this grain refinement: at hypoeutectic levels (0.5 wt.% Sc), coarse elongated grains (~300 µm) dominate, but increasing to 0.7 wt.% Sc reduces the grain size and the grain structure transformed to equiaxed morphology. Beyond 0.7 wt.%, further Sc additions yield diminishing returns, establishing a "modification threshold" of ~0.6-0.7 wt.% for binary Al-Sc cast alloy [42, 45]. Additions of Zr or Ti further improve refining effect by reducing lattice mismatch (e.g., $Al_3(Sc, Ti)$ at 300 °C), enhancing nucleation efficiency and coarsening resistance, which delays over-aging up to 300-350 °C [44, 46].

Sc additions promote significant precipitation hardening in cast Al alloys, with secondary Al_3Sc-type precipitates forming during aging at ~300-350 °C (higher than typical Mg/Zn/Cu precipitates at 120-200 °C) [36, 43, 47]. These coherent precipitates disperse in the Al matrix, exerting Zener-drag on grain boundaries to prevent recrystallization and stabilize microstructures under high temperatures or deformation. For example, ~0.38 wt.% Sc increases yield strength (YS) from ~15 MPa in pure cast Al to 199 MPa after aging at 288°C for 8 hours [35].

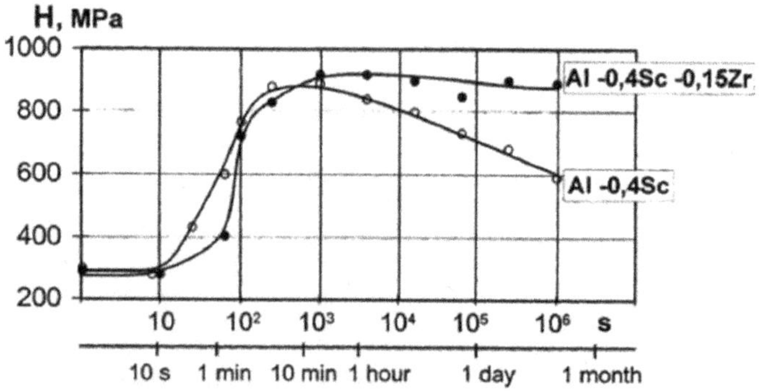

Figure 2.12 Microhardness of Al-0.4 wt.% Sc and Al-0.4 wt.% Sc-0.15 wt.% Zr alloy during isothermal holding at 350°C [48].

Figure 2.12 shows that the Al-0.4 wt.% Sc alloy containing Zr retains high strength for a longer duration than the Zr-free alloy due to the formation of $Al_3(Sc_{1-x}Zr_x)$ precipitates. The lower diffusivity of Zr in Al enhances resistance to coarsening, maintaining thermodynamic stability. In multicomponent casts, these precipitates may incorporate Mg or Mn, deviating from ideal stoichiometry but still contributing to overall hardening [48, 49].

2.9.2 Thermodynamics of Zr + Ti addition

Zirconium (Zr) and titanium (Ti) additions to Al alloys are primarily governed by the peritectic reaction of the Al-Zr and Al-Ti binary phase diagrams (Figure 2.13), which influence solute solubility, supersaturation, precipitate formation, and thermal stability [12]. In the Al-Zr system, Zr exhibits a maximum solid solubility of ~0.28 wt.% at the peritectic temperature of 660.8°C, decreasing sharply with temperature, leading to metastable supersaturation during non-equilibrium solidification [50]. This drives the precipitation of nanometer-scale Al_3Zr with a metastable cubic $L1_2$ structure during aging at 375-425°C, which transforms sluggishly to the equilibrium tetragonal $D0_{23}$ phase above 475°C due to high activation energies for nucleation and growth, influenced by Zr's low diffusivity in Al (D ~ 10^{-14}-10^{-13} m^2/s at 475°C). The stability of the $L1_2$ phase stems from its small lattice parameter mismatch with α-Al, which lowers interfacial energy and sustains coherency up to radii of ~35 nm, while slow Zr diffusion (activation energy Q ≈ 242 kJ/mol) retards coarsening, enhancing creep resistance and thermal stability up to 0.75 T_m (melting point of Al) [51, 52].

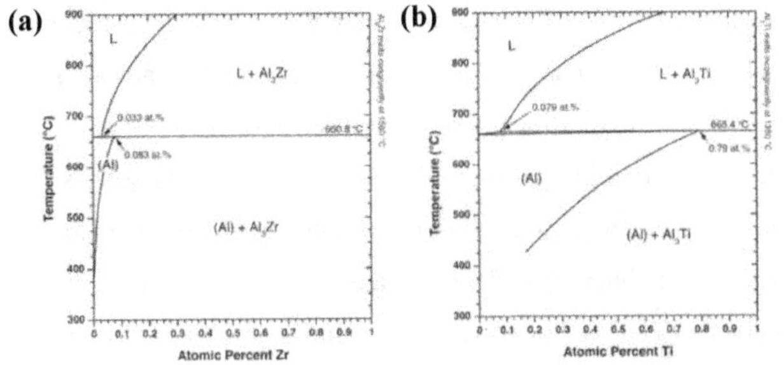

Figure 2.13 Equilibrium phase diagram of (a) Al-Zr and (b) Al-Ti [12].

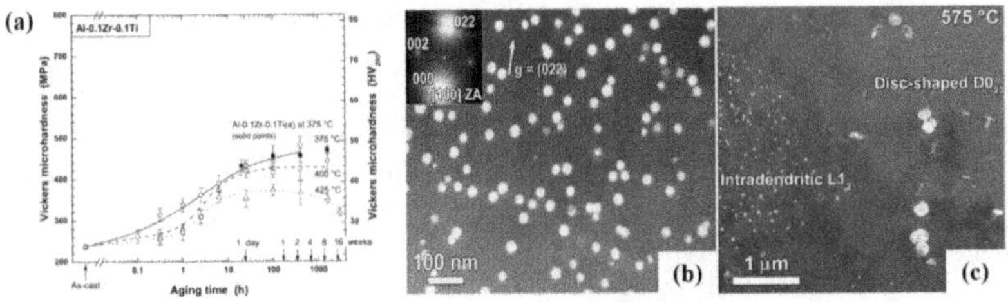

Figure 2.14 (a) Vickers microhardness of Al-0.1Zr-0.1Ti alloy during isochronal aging at 375°C, 400°C and 425°C; (b-c) dark-field TEM images of Al-0.1Zr-0.1Ti after isochronal aging at 425°C (b) and 575°C (c) [51, 52].

Ti addition reduces the metastable Al-Zr solvus, increasing effective Zr supersaturation and accelerating $L1_2$ nucleation kinetics, but does not improve coarsening resistance or delay the $L1_2 \rightarrow D0_{23}$ transformation, as observed in Al-0.1Zr-0.1Ti alloys where precipitate radii remain identical (~11-12 nm after 1600 h at 425°C) to binary Al-Zr counterparts (Figure 2.(a-b)). Ti partitions weakly to $Al_3(Zr_{1-x}\ Ti_x)$ (x ≈ 0.17 at 425°C), reducing the lattice mismatch and stabilizing the $L1_2$ phase during overaging [51]. However, the thermodynamics promote heterogeneous transformation on dislocations due to large $D0_{23}$ mismatches, leading to disk-shaped precipitates (~200 nm diameter, 50 nm thick) with cube-on-cube orientation (Figure 2.c). The combination of Zr and Ti enhances precipitation hardening through increased supersaturation and refined microstructures (equiaxed grains from primary Al_3(Zr, Ti) nucleants). However, their thermodynamic benefits are limited by diffusion-controlled coarsening and phase stability, resulting in no significant delay in overaging above 450°C compared to binary systems [52].

2.10 Aerospace grade alloys

2.10.1 Thermal stability of Al-Cu alloys

Al-Cu alloys typically contain 3-11% Cu, which provides excellent heat-treatable strengthening and thermal stability. These alloys are commonly used in components exposed to temperatures up to ~150°C, particularly in applications with relatively simple geometries, such as fasteners, brackets, and fittings in areas exposed to engine heat or aerodynamic heating [53]. The microstructure of binary Al-Cu alloys consists of α-Al solid solution and Al_2Cu intermetallic phases in the as-cast condition. At 4-5% Cu, the eutectic structure forms a continuous network, while increasing the Cu content to 10% transforms it into a denser, fully interconnected network. The precipitation sequence during aging of Al-Cu alloys is described as follows: solute clusters → GP zones → θ″ → θ′ → θ. Figure 2.15a shows the Al-Cu phase diagram and solvus lines for various metastable precipitates [54]. In engineering practice, the solutionized alloy is often artificially aged by heating within the two-phase region, promoting the formation of specific metastable precipitates depending on time and temperature. At low aging temperatures or during natural aging, Cu atoms cluster within the supersaturated matrix, forming single-layer GP-I zones (Guinier-Preston zones) parallel to the (001) plane of α-Al. With further aging above the GP zone solvus (Figure 2.15a), θ" (GP-II) precipitates form by Cu diffusion from GP-I zones. θ" (Al_3Cu) has an ordered tetragonal structure (a = b = 4.04 Å, c = 7.68 Å) with two Cu layers separated by three Al layers (Figure 2.15b), and is fully coherent with the α-Al matrix, though misfit strain occurs along $(100)_{Al}$ and $(010)_{Al}$ [55, 56].

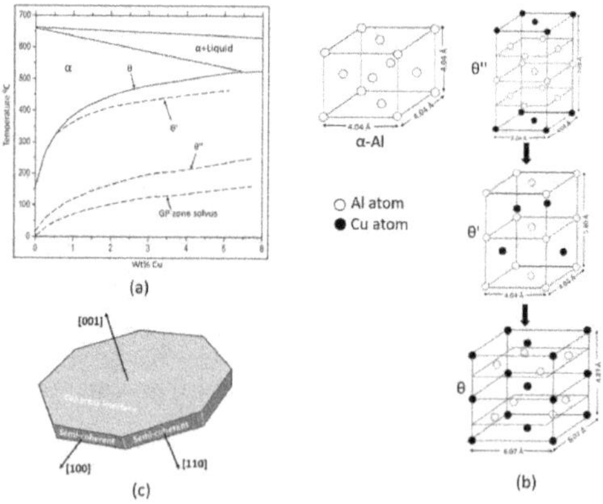

Figure 2.15 (a) Al-Cu binary phase diagram; (b) crystal structures of α-Al and different metastable (θ″, θ′) and stable (θ) precipitates; and (c) θ′ precipitate morphology and interfaces [54].

As the aging process continues, θ" precipitates transform into θ′, which has a body-centered tetragonal structure (a = b = 4.04 Å, c = 5.80 Å, space group I_4/mmm) with three Al and two Cu layers (Figure 2.15b). θ′ nucleates at θ" sites and grows by further Cu diffusion, maintaining the same orientation with α-Al: $(001)_{\theta'} \parallel (001)_{Al}$ and $[100]_{\theta'} \parallel [100]_{Al}$ [57]. θ′ forms as plates parallel to $(010)_{Al}$, remaining coherent along this plane but becoming semi-coherent or incoherent along $(100)_{Al}$ and $(010)_{Al}$. The semi-coherent sides of the plates are octagonal, faceted along $[100]_{\theta'}$ and $[110]_{\theta'}$ (Figure 2.15c). The $\{110\}_{\theta'}$ interfaces act as diffusion pathways for Cu, which promotes coarsening of θ′ precipitates during prolonged thermal exposure; ledge dislocations at these interfaces also facilitate the incorporation of Cu atoms. The equilibrium θ precipitate (Al_2Cu) is tetragonal (a = b = 6.07 Å, c = 4.87 Å, space group I_4/mcm; Figure 2.15b) and incoherent with α-Al. It forms from θ′ through a similar transformation, with its habit plane parallel to $(100)_{\alpha\text{-}Al}$ [58]. To slow the coarsening of θ′ into θ and improve the microstructural stability, alloying elements such as Mn, Ni, Sc, Zr, and Ti are incorporated. These elements refine the microstructure and improve mechanical strength for high-temperature applications.

The addition of Mn in (0.5-0.7%) to Al-Cu alloys enhances their room and elevated temperature strength during natural aging while reducing susceptibility to weld cracking. This improvement stems from the formation of thermally stable Al_6Mn and T-$Al_{20}Cu_2Mn_3$ phases, which significantly enhances the material heat resistance [59]. Further incorporation of Ni to Al-Cu-Mn alloys significantly enhances their elevated-temperature stability by forming thermally resistant intermetallic phases such as Al_7Cu_4Ni, Al_3CuNi, Al_3Ni_2, and $Al_3(Cu, Ni)_2$. These intermetallics exhibit complex crystal structures (e.g., rhombohedral Al_7Cu_4Ni and hexagonal Al_3Ni_2), remain stable at temperatures up to 400°C, mitigating strength degradation. Lin et al. [60] investigated how Ni additions (0.5-1.5 wt%) influence the high-temperature performance of squeeze-cast Al-Cu-Mn-Fe alloys. Their study revealed a direct correlation between Ni content and the volume fraction of thermally stable intermetallic phases, including Al_9FeNi, Al_3CuNi, and T-$Al_{20}Cu_2Mn_3$. As Ni concentration increased, these precipitates showed enhanced stability at 300°C, leading to significant improvements in mechanical properties under elevated temperature [60, 61]. Table 2.1 summarizes the main precipitate phases and their crystal structures in Al-Cu alloys.

Table 2.1 Crystallographic parameters of main strengthening phases in Al-Cu alloy [52, 54, 62-68].

Phase	Crystal System	Lattice Parameters (Å)			Orientation	Precipitation Temperature (°C)
		a	b	c		
θ-Al_2Cu	Tetragonal	6.07	6.07	4.87	$(100)_{\theta} \parallel (100)_{Al}$; $[010]_{\theta} \parallel [010]_{Al}$	300–350
θ'-Al_2Cu	Tetragonal	4.04	4.04	5.80	$(001)_{\theta'} \parallel (001)_{Al}$; $[100]_{\theta'} \parallel [100]_{Al}$	150-200
θ''-Al_3Cu	Tetragonal	4.04	4.04	7.68	$(001)_{\theta''} \parallel (001)_{Al}$; $[100]_{\theta''} \parallel [100]_{Al}$	100-200
T-$Al_{20}Cu_2Mn_3$	orthorhombic	2.42	1.55	0.775	$[010]_T \parallel \langle 010 \rangle_{Al}$, $\{\bar{7}03\}_T \parallel \{002\}_{Al}$; $[010]_T \parallel \langle 010 \rangle_{Al}$, $\{\bar{1}\bar{1}01\}_T \parallel \{002\}_{Al}$; $[010]_T \parallel \langle 010 \rangle_{Al}$, $\{004\}_T \parallel \{002\}_{Al}$	300-450
Al_6Mn	orthorhombic	6.49	7.55	8.87	Semi-coherent; $[100]_{Al6Mn} \parallel [100]_{Al}$,	300-500
Al_7Cu_4Ni	orthorhombic	4.01	8.12	14.81	Semi-coherent; often on {111}Al planes.	400-500
Al_3CuNi	Hexagonal	4.06	4.06	4.90	Incoherent	400-550
Al_3Ni_2	Hexagonal	3.99	3.99	4.88	Incoherent	400-600
Al_3Sc	Cubic	4.10	4.10	4.10	Coherent; OR: $(001)_{L12} \parallel (001)_{Al}$, $[100]_{L12} \parallel [100]_{Al}$	250-350
Al_3Zr	Cubic ($L1_2$)	4.08	4.08	4.08	Coherent; OR: $(001)_{L12} \parallel (001)_{Al}$, $[100]_{L12} \parallel [100]_{Al}$	350-450 (metastable)
	Tetragonal (DO_{23})	4.01	4.01	17.3	Semi-coherent	400-500 (stable)
Al_3Ti	$L1_2$	4.05	4.05	4.05	Coherent; OR: $(001)_{L12} \parallel (001)_{Al}$, $[100]_{L12} \parallel [100]_{Al}$	300-400 (metastable)
	Tetragonal (DO_{23})	3.84	3.84	33.5	Semi-coherent	450-550

Another effective strategy to enhance the thermal stability of Al-Cu alloys is microalloying with secondary elements that stabilize the key strengthening metastable θ′ precipitates. The addition of Zr in Al-Cu-Mn alloys significantly enhances their high-temperature stability. Early studies by Shyam et al. [69] compared Al-Cu-Mn-Zr alloys with conventional Al-Cu (206) and

Al-Si-Cu (319) alloys, revealing that while the base alloys initially displayed better mechanical properties at room temperature, the Al-Cu-Mn-Zr alloys outperformed them after prolonged exposure to 300°C. For example, the Al-5CuNiMnZr alloy exhibits a yield stress of ~250 MPa at room temperature, which is lower than the Al-5CuMg alloy (about 450 MPa). However, after exposure to 300°C, the Al-5CuNiMnZr alloy maintains a strength of ~130 MPa, while the Al-5CuMg alloy drops to about 60 MPa (Figure 2.16). This improvement is attributed to the ability of θ′ precipitates in Al-Cu-Mn-Zr alloys to maintain their morphology and resist coarsening, unlike in the base alloys where θ′ transforms into the stable θ phase, leading to property degradation. Further research by Bahl et al. [70] showed that these alloys retain their mechanical strength even after 5000 hours exposure at 300°C, with only a minor reduction in θ′ precipitate density.

To investigate the individual and combined effects of Mn and Zr microalloying on the thermal stability of Al-Cu-Mn-Zr alloys, Poplawsky et al. [71] conducted a comprehensive study using different alloy systems, including Al-Cu-Mn (ACM), Al-Cu-Zr (ACZ), and the base Al-Cu-Mn-Zr alloy. Their findings show that Al-Cu-Mn alloy maintains its mechanical strength at room temperature even after 200 hours at 300°C, while Al-Cu-Zr alloys remain stable only up to 200 °C. In contrast, the Al-Cu-Mn-Zr alloy demonstrates stability up to 350 °C. The observed Mn segregation pattern in Al-Cu-Mn-Zr alloys up to 300°C aligns with earlier results by Shyam et al. [69], with Mn segregation observed at semi-coherent interfaces and only minor segregation at coherent interfaces.

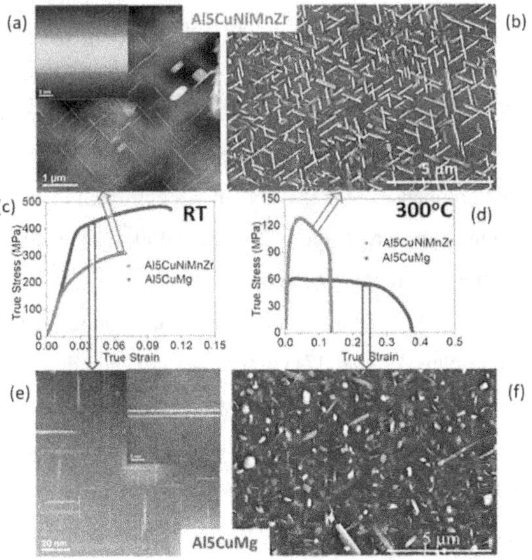

Figure 2.16 (a) and (b) show the microstructures of the Al-Cu-Mn-Zr alloy in the peak-aged condition and after 200 hours of thermal exposure at 300°C, respectively; (c) and (d) true stress-true strain curves from tensile tests conducted at room temperature and 300°C; (e) and (f) display the microstructures of the conventional Al-Cu-Mg alloy under similar thermal conditions [69].

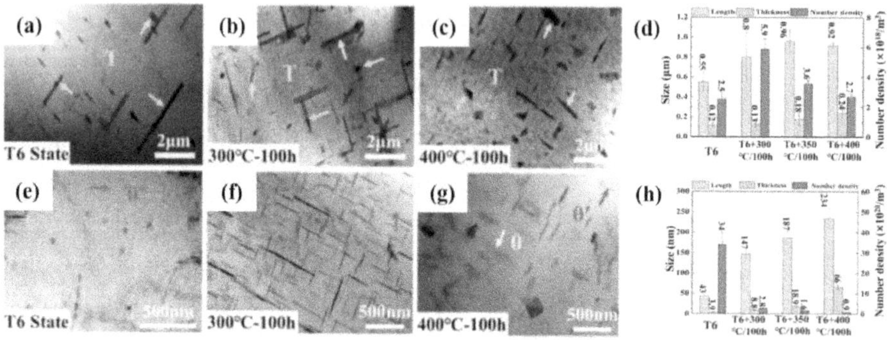

Figure 2.17 TEM images of the T and θ′ phases in the Al-8.4Cu-2.3Ce-1.0Mn-0.5Ni-0.2Zr alloy (T6) after 100 h of aging at different temperatures: (a, e) T6 state; (b, f) 300°C; (c, g) 400 °C. (d, h) distributions of length, thickness, and areal density for the T and θ′ phases [72].

Xiang et al. [72] reported that the T-$Al_{20}Cu_2Mn_3$ phase re-precipitates after 300°C for 100 hours thermal exposure which enhances the thermal stability of Al-Cu-Ce-Mn-Ni-Zr alloy. The area density of the T phase initially increases and then reduces as the re-aging temperature increases, with no significant coarsening observed until 400°C. Figure 2.17 illustrates the proportion of T phase particles thinner than 100 nm increases significantly, and the area density increases from 2.5×10^{18} to 5.9×10^{18} m^{-3}. Additionally, Ni, Ce, and Zr are found to segregate at the T phase/matrix interface, which further enhances the thermal stability of the T phase. Notably, the T phase also helps compensate for the loss in strength caused by the coarsening of θ′ precipitates after thermal exposure (Figure 2.17). At 400 °C, the alloy achieves a tensile strength of 148 MPa while maintaining 9.7% ductility.

Minor Ti additions improve Al-Cu-Mn-Zr alloy high-temperature stability by forming $L1_2$-Al_3Ti precipitates [73]. Poplawsky et al. [71] observed that Ti addition leads to $L1_2$-$Al_3(Zr_xTi_{1-x})$ precipitates at α-Al/θ′ interfaces. These precipitates are coherent with both $(100)_{Al}$ and $(001)_{\theta'}$ planes, stabilizing θ′ and reducing misfit strain. This $L1_2$ formation also inhibits θ′ thickening along coherent interfaces by "ledge poisoning" [74]. Semi-coherent interfaces are depleted of these precipitates, supported by DFT calculations [71, 74], due to the energetic favorability of the coherent θ′/ $Al_3(Zr_xTi_{1-x})$/α-Al structure. The self-diffusion coefficients ($D_{Mn} > D_{Zr} > D_{Ti}$) govern the segregation [74]. Mn diffuses fastest to both coherent and semi-coherent θ′ interfaces, while Zr primarily segregates to coherent interfaces. During thermal exposure, Mn initially segregates broadly, then Zr concentrates at coherent interfaces. Further heat treatment allows Ti to form $Al_3(Zr_xTi_{1-x})$ precipitates, mainly at coherent interfaces. The segregation sequence is schematically illustrated in Figure 2.18.

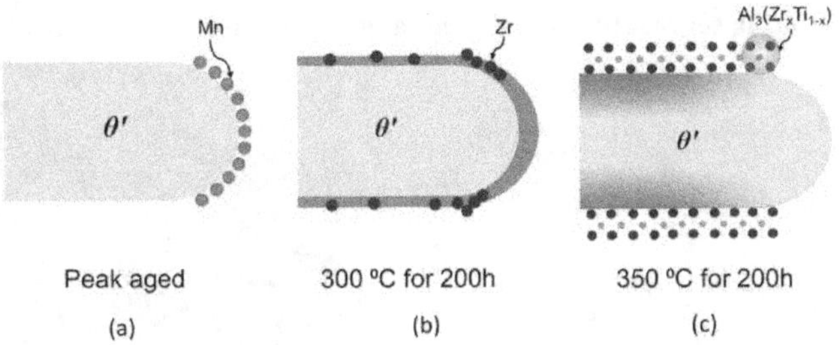

Figure 2.18 Schematics illustrate: (a) Mn enrichment at the semi-coherent interface of θ′ precipitates in the peak-aged state; (b) Zr accumulation following extended thermal exposure at 300°C for 200 h; and (c) the emergence of $Al_3(Zr_xTi_{1-x})$ precipitates at the boundaries of coherent and semi-coherent interfaces, along with Mn diffusion into the θ′ precipitate after heat treatment at 350 °C for 200 h [54].

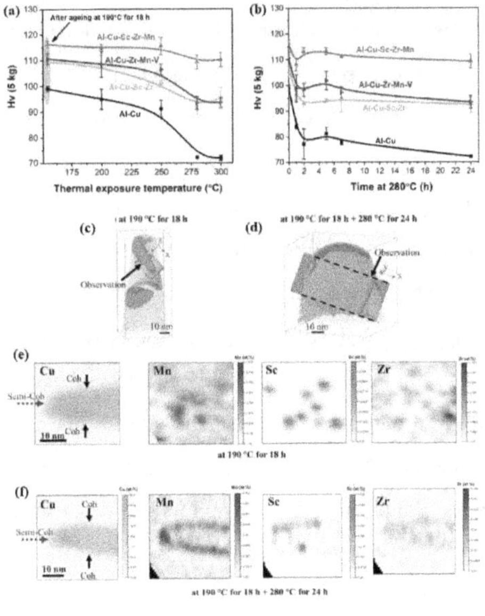

Figure 2.19 The hardness of different micro-alloyed Al-Cu alloys following 24-hour thermal treatments at a range of temperatures; (b) The progression of hardness in artificially aged alloys subjected to thermal exposure at 280°C; (c, d) Three-dimensional atom maps of Al (cyan) and Cu (brown) for Al-Cu-Sc-Zr-Mn alloy after 18 h at 190°C and after 24 h at 280°C, with θ' precipitates delineated by an 8.0 at.% Cu iso-concentration surface; (e, f) 2D cross-sectional contour plots of θ' precipitates, indicating segregation at coherent and semi-coherent interfaces [77].

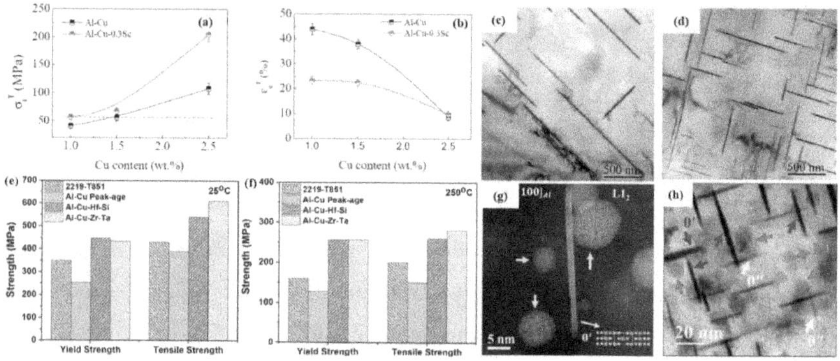

Figure 2.20 (a, b) The tensile strength (σ_t^T) and elongation to fracture (ε_t^T) of the alloys tested at 300°C, as a function of varying Cu concentration; TEM images illustrating the size and distribution of θ′ precipitates in (c) Al-2.5Cu ; and (d) Al-2.5Cu-0.3Sc alloy; Tensile properties of different microalloyed Al-Cu alloys tested at (e) 25°C; (f) 250°C; (g) θ′ plates formed on (Al, Si)$_3$Hf precipitates in an Al-2Cu-0.3Hf-0.3Si (at.%) alloy; (h) Centered bright-field TEM image showing both θ″ and θ′ phases nucleating on pre-existing L1$_2$-ordered precipitates in Al-1.48Cu-0.24Zr-0.18Ta alloy (at.%) [80-82].

Similarly, microalloying with 0.18 at.% Sc leads to the formation of thermally stable Al$_3$Sc dispersoids, with Sc atoms preferentially segregating at the α-Al/θ′ interfaces [75]. This co-stabilization of both θ′ and Al$_3$Sc phases significantly improves the high-temperature properties of Al-Cu alloys [76]. Figure 2.19 (a-b) shows that the Al-Cu-Sc-Zr-Mn alloy exhibits the highest thermal stability, maintaining a hardness above 115 HV even after exposure to 280°C, while the base Al-Cu alloy drops below 80 HV. The atom probe maps and 2D contour plots (Figure 2.19 (c-f)) indicate that enhanced stability is due to the segregation of Mn, Sc, and Zr at θ′ precipitate interfaces, which inhibits coarsening and retains hardness during prolonged thermal treatment [77]. In Al-Cu-Mg systems, trace additions of Mg (0.09-0.13 wt.%) accelerates θ′ precipitation, enabling the alloy to maintain thermal stability at 300°C for up to 1000 hours [78]. In addition, the high-temperature performance of AA2219 alloy is significantly enhanced by microalloying with 0.8 wt% Sc, 0.45 wt% Mg, and 0.2 wt% Zr, which promotes grain refinement and precipitation of thermally stable phases such as Al_3Sc, Al_3Zr, and Ω, alongside conventional strengthening precipitates of θ′ and θ″ [79]. Figure 2.20(a-b) illustrates that increasing Cu concentration (from 1.5 to 2.5 wt.%) in Sc-modified alloys enhances tensile strength with the Al-2.5Cu-0.35Sc alloy achieving peak values of ~205 MPa in tensile strength and ~10% elongation at 300°C. TEM images (Figure 2.20(c-d)) reveal that Sc additions promote the formation of fine, ordered θ′ precipitates and $L1_2$-Al_3Sc dispersoids, inhibiting dislocation glide and retaining strength up to 300°C [80]. The addition of 0.3 at.% Hf and 0.3 at.% Si to the Al-2Cu alloy (Figure 2.20(e-f)) facilitates the nucleation of θ′ plates on $(Al, Si)_3Hf$ dispersoids (Figure 2.20g), leading to good high-temperature yield strength (~256 MPa at 250°C) by reducing interfacial energy and suppressing precipitate coarsening [81]. Similarly, Zr and Ta additions retain strength of ~257 MPa at 250°C by segregating at θ′ interfaces (Figure 2.20h) [82]. The formation of a thermally stable Ω phase in the Al-Cu system, achieved through Sc alloying and a two-stage heat treatment, forms an AlCuMgAgSc precipitate with Mg/Ag-rich outer layers and Cu/Sc-rich cores [83]. As a result of this microstructural feature, the alloy maintains a tensile strength of 100 MPa at 400°C. Subsequent investigations on Al-Cu-Sc-Fe-Si alloys [84, 85] indicate that Sc-Fe-Si segregation at θ′ interfaces significantly enhance creep resistance, with the alloy sustaining over 1000 hours at 300°C under a stress of 30 MPa. These compositional modifications and their effects on room and elevated temperature mechanical properties are summarized in Tables 2.2a and 2.2b.

Table 2.2a Compositions of widely used commercial Al-Cu based alloys in wt.%.

Alloy	Alloying Elements															Ref.
	Cu	Mn	Mg	Fe	Si	Zr	Ti	Co	Sb	Sc	Ni	Ta	Hf	V	Ag	Al	
2024	4.4	0.6	1.5	≤0.5	≤0.5	-	-	-	-	-	-	-	-	-	-	Bal.	[86]
2219	6.3	0.3	-	≤0.3	≤0.2	0.18	0.03	-	-	-	-	-	-	0.06	-	Bal.	[79]
2618	2.7	-	1.8	1.3	0.25	-	-	-	-	-	1.2	-	-	-	-	Bal.	[87]
2618SZ	2.23	-	1.2	1	-	0.3	-	-	-	0.3	1.1	-	-	-	-	Bal.	[88]
2219SZ	6.3	0.3	-	0.13	0.05	0.34	0.03	-	-	0.4	-	-	-	0.06	-	Bal.	[79]
RR350	5.02	0.2	-	0.09	0.03	0.17	0.21	0.25	0.16	-	1.5	-	-	-	-	Bal.	[69]
ACMN	4.7	0.99	0.0009	0.01	-	-	-	-	-	-	0.49	-	-	-	-	Bal.	[61]
A201	5.2	0.5	-	0.15	0.1	-	0.35	-	-	-	-	-	-	-	0.5	Bal.	[89]
ACMFN	4	-	2	1.1	-	-	-	-	-	-	1.1	-	-	-	-	Bal.	[90]
ACMN	4.9	1.02	0.009	0.12	-	-	-	-	-	-	1.48	-	-	-	-	Bal.	[61]
ACHS	4.48	-	-	-	0.3	-	-	-	-	-	-	-	1.89	-	-	Bal.	[81]
ACZT	3.36	-	-	-	-	0.78	-	-	-	-	-	1.17	-	-	-	Bal.	[82]
ACMg	5.18	-	0.37	-	0.14	-	0.02	-	-	-	-	-	-	-	-	Bal.	[69]
ACMFN	5.5	0.6	-	1	-	-	-	-	-	-	1.5	-	-	-	-	Bal.	[91]

Table 2.2b Ambient and elevated temperature mechanical properties of Al-Cu based alloys.

Alloy	Heat-treatment	Room Temp Tensile Strength (MPa)	Room Temp Yield Strength (MPa)	Room-Temp Elongation (%)	High Temp	High-Temp Tensile Strength (MPa)	High-Temp Yield Strength (MPa)	High-Temp Elongation (%)	Ref.
2024	T3, T4, T6, T8	~510	~430	~20	200	~180	~150	~10	[86]
2219	T6, T62, T81	~455	~345	~12	250	~200	~160	~21	[79]
2618	T6, T7	~330	~256	~5	300	~98	~90	~20	[87]
2618SZ	T6, T7	~390	~340	~6	300	~170	~140	~21	[88]
2219SZ	T6	~536	~456	~5	250	~231	~227	~5	[79]
RR350	T6	~310	~220	~6	300	~130	~120	15	[69]
ACMN	T6	~290	~170	~4.6	300	~141	~117	~4.7	[61]
A201	T7	~450	~420	~6	300	~175	-	~9	[89]
ACMFN	T6	~378	-	~5.6	200	~305	-	~7.84	[90]
ACMN	T6	~205	~114	~3.4	300	~105	~86	~10.2	[61]
ACHS	T6	~540	~448	~7	250	~260	~256	-	[81]
ACZT	T6	~610	~435	~4	250	~280	~257	-	[82]
ACMg	T6	~480	~380	~11	300	~60	~54	~35	[69]
ACMFN	T7	~285	~256	~8	300	~156	~145	~23	[91]

2.10.2 Thermal stability of Al-Zn-Mg-Cu alloys

The 7xxx series Al alloys are a class of heat-treatable Al alloys containing Zn as the principal alloying element, typically accompanied by Mg and Cu. The maximum solubility of Zn in Al is 31.6 wt.% at 275°C, which decreases to 5.6 wt.% at 125°C. While the addition of Zn alone has minimal strengthening effects, combining it with Mg leads to the formation of strengthening precipitates, such as $MgZn_2$ (η) and $Al_2Mg_3Zn_3$(T) during aging. These precipitates are responsible for the significant enhancement in mechanical properties observed in these alloys. The introduction of small amounts of Cu further accelerates the precipitation process, increasing the density of these hardening phases at the early stages of aging [92-94]. Figure 2.21a illustrates the equilibrium phase composition in Al-rich regions of 7xxx series Al alloys changes with temperature. The phase diagram reveals that increasing Cu content increases the formation of the η and S (Al_2CuMg) phase. At lower Cu levels, the Zn/Mg ratio becomes the dominant factor influencing phase composition. When Zn/Mg ratio is very low, the microstructure is primarily composed of α + S + T phases, or medium α + T phases without the η phase. As the Zn/Mg ratio increases, the η precipitate starts to form and its phase fraction becomes more pronounced, often alongside S and T phases [93, 95].

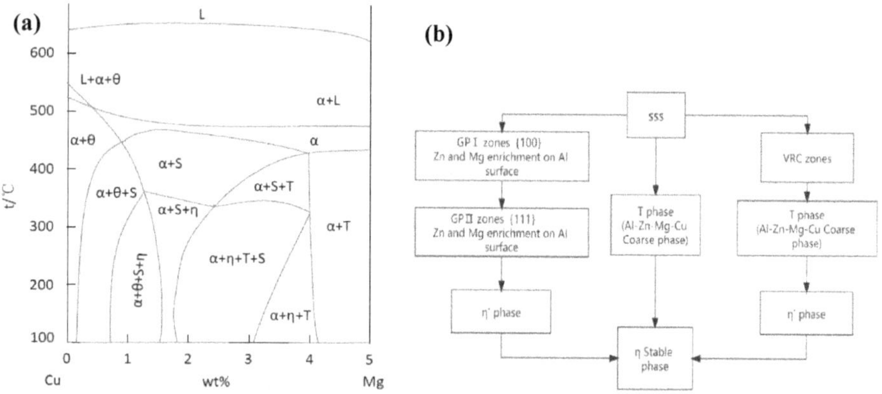

Figure 2.21 (a) Variable temperature section (isopleth) of the aluminum-rich corner of the equilibrium phase diagram for 7xxx series aluminum alloys (with Al as the balance ~90-95 wt.%); and (b) Schematic illustration of the precipitation sequence in 7xxx series Al alloys [93, 95].

The precipitation sequence in these alloys can follow three main pathways: (1) supersaturated solid solution (SSS) transforms through GP zones (GP I and GP II) to η' and then η precipitates; (2) with high Mg content, SSS at defects and grain boundaries leads to T phase and coarse Al-Zn-Mg-Cu phases before forming η; and (3) in vacancy-rich regions, SSS forms vacancy-related clusters, then T phase and coarse Al-Zn-Mg-Cu phases, and finally η precipitate. All three routes end at the stable η precipitate, as shown in Figure 2.21b [95]. The principal strengthening phases in Al-Zn-Mg-Cu alloys are the GP zones and the η' precipitate. GP zones are fully coherent with the Al matrix and appear as either spherical (GP I) or strip-like (GP II) regions enriched in Mg and Zn. The η' precipitate is a metastable, semi-coherent hexagonal plate shape, providing peak age-hardening. The equilibrium η precipitate is disc-shaped and incoherent with the matrix [96]. Under certain compositions, the T phase may replace the η precipitate, and the presence of Cu can lead to S phase formation at grain boundaries, which may promote pitting corrosion and reduce fracture toughness. Table 2.3 provides an overview of the main precipitate phases and their crystal structural characteristics in Al-Zn-Mg-Cu alloys.

Table 2.3 Crystallographic parameters of main strengthening phases in Al-Zn-Mg-Cu alloy [93, 97-100].

Phase	Crystal System	Lattice Parameters (Å)			Orientation	Precipitation Temperature (°C)
		a	b	c		
η'-$MgZn_2$	Hexagonal	0.496	-	1.403	semi-Coherent, $(0001)_{\eta'}$ //$\{111\}_{Al}$; $(11{-}20)_{\eta'}$//$<112>_{Al}$; $(10{-}10)_{\eta'}$//$(110)_{Al}$; habit plane $(111)_{Al}$	120-250
η-$MgZn_2$	DO_{23}	0.516-0.522	-	0.849-0.855	Incoherent; 13 OR; classic OR $(0001)_{\eta}$//$\{111\}_{Al}$, $(10{-}10)_{\eta}$//$\{110\}_{Al}$	150-476
T' ($Al_2Mg_2Zn_3$)	Cubic	1.422	-	-	Metastable; Semicoherent; Equiaxed Polytopic	471-476
T ($Al_2Mg_2Zn_3$)	Cubic	1.456	-	-	Equilibrium; Incoherent, $(060)_T$//$(111)_{Al}$; $(103)_T$//$(2{-}20)_{Al}$	174
S' (Al_2CuMg)	Orthorhombic	0.401	0.926	0.7124	Incoherent; Bar shape	150-300
S (Al_2CuMg)	Orthorhombic	0.4	0.025	0.715	semi-Coherent	400-494

The 7xxx series alloys are recognized for their ultra-high strength, with tensile strengths often surpassing 500 MPa at ambient temperature, making them some of the most robust Al materials. Figure 2.22a provides a comparison of the mechanical properties of different 7xxx and 2xxx series alloys, highlighting the higher yield strength but lower ductility of the 7xxx series alloys. While 7050-T7 and 7075-T6 alloys exhibit lower fracture toughness compared to 2024-T3, the 7475-T3 alloy stands out with a significantly higher toughness. Due to their excellent strength at ambient temperature, 7xxx series alloys are predominantly used in applications demanding high load-bearing capacity, such as upper wing surfaces, fuselage stringers, and structural frameworks in aerospace engineering. Figure 2.22(b-d) illustrate the application of these alloys in aircraft wings and floor beams, underscoring their importance in critical structural components [8, 101-103]. However, as modern industries continue to advance, there is a persistent and growing need for structural materials that can maintain their properties at elevated temperatures. For example, the outer surfaces of high-speed spacecraft can reach temperatures up to 215°C at speeds above Mach 2.5. Despite their excellent strength at lower temperatures compared to 2xxx series aerospace grade alloys (2618-T6 and 2219-T6 alloys), Al-Zn-Mg-Cu alloys lose much of their mechanical integrity after high temperature exposure. The 7075 alloy retains only about 30% of its room-temperature tensile strength at 200 °C and only 10% at 300 °C, which limits their use to below 150 °C [104]. This highlights the ongoing need to develop an alloy that combines the excellent room temperature properties of 7075-T6 with the high-temperature performance of 2618-T6 or 2219-T6 alloy.

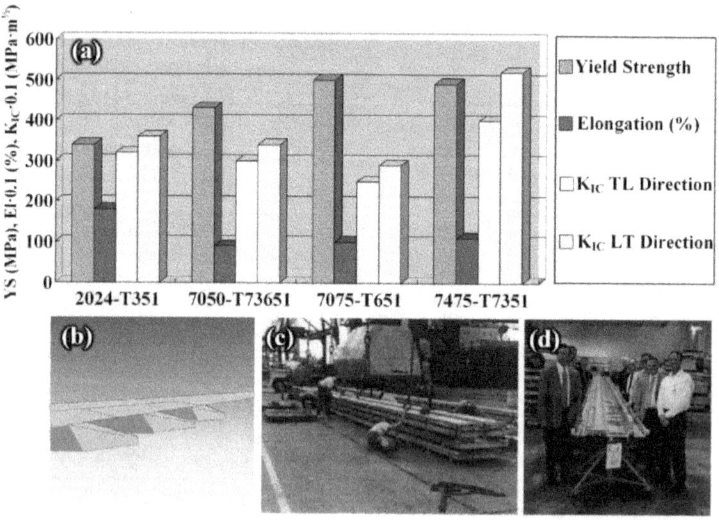

Figure 2.22 (a) Mechanical properties of selected aerospace Al alloys; (b) 7075 alloy in wing structures; (c, d) floor beam plates and a machined cross-section from a floor beam [8].

To address this challenge, researchers have investigated the thermal stability of these alloys. Dai et al. examined T7452-treated 7085 Al alloy and found that exposure to 100°C and 175°C led to the transformation of the η′ precipitate to the η precipitate, resulting in larger and more widely spaced precipitates and a decrease in strength [105]. Zhao et al. [106] discovered a four-stage aging treatment process, including a natural aging step before re-aging, which improved the formation of GP zones and enhanced both strength and thermal stability at 120°C. These findings highlight that the rapid transformation and coarsening of η′ to η precipitates are key factors in the loss of mechanical properties at elevated temperatures. The η′ precipitates tend to coarsen quickly because of the high diffusivity of Zn and Mg in Al, reducing their effectiveness. Strategies to increase the density of η′ precipitates can slow this decline, but rapid coarsening remains a challenge. In addition to η′/η precipitates, T′/T phases also form in these alloys and offer better heat resistance, though generally at the expense of some strength [107, 108]. For example, a T-phase strengthened Al-Zn-Mg alloy designed by Takata et al. [109] achieved a yield strength of ~ 260 MPa at 200 °C, attributed to the T phase's stability (Figure 2.23). The co-existence of T′/T and η′/η precipitates, with a low Zn/Mg ratio, may offer a balance of strength and heat resistance. Zou et al. [108] have shown that T′ phases grow more slowly than η′ precipitates at elevated temperatures, suggesting that a microstructure containing both could improve thermal stability by influencing precipitate coarsening behavior. Similarly, the addition of 4 wt.% Ni to the Al4-Mg-7.8Zn-4Cu alloy achieves a yield strength of ~350 MPa and a tensile strength of ~420 MPa at room temperature. This alloy retains significant strength at 200°C, exhibiting a yield strength of ~220 MPa and a tensile strength of ~260 MPa. This is attributed to the presence of stable T, Al_3Ni, and $Al_3(Ni, Cu)_2$ intermetallic phases within the α-Al matrix [87].

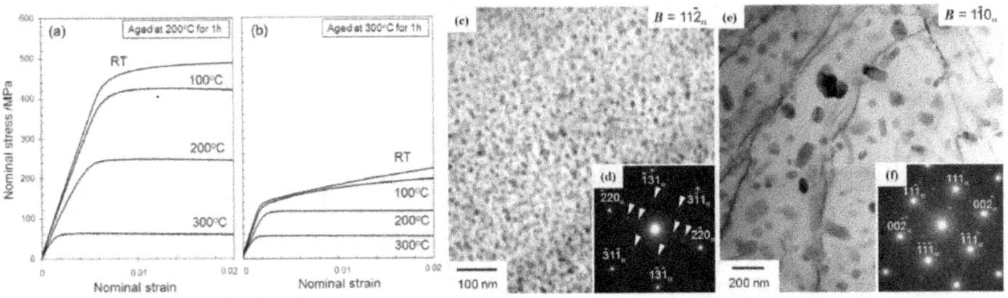

Figure 2.23 Nominal stress-strain curves for Al-5Mg-3.5Zn alloy tested at different temperatures: (a) aged at 200 °C for 1 h, (b) aged at 300 °C for 1 h. (c, e) TEM bright-field images showing precipitates within the α-Al matrix. (d, f) Corresponding selected area diffraction patterns for the alloy aged at (c, d) 200 °C and (e, f) 300 °C for 1 h [109].

One promising approach to enhance the elevated temperature strength of Al-Zn-Mg-Cu alloys is the addition of Sc. Several studies have demonstrated that Sc additions, combined with Zr, has a significant impact on both thermal stability and mechanical properties [104]. The presence of Sc and Zr leads to the formation of core-shell $Al_3(Sc, Zr)$ particles during solidification, which refine the grain structure. The combined addition of 0.15 wt.% Sc and 0.14 wt.% Zr strengthens grain boundaries and increases yield strength from ~577 MPa to ~605 MPa of 7055-T6 alloy [110]. However, excessive Sc and Zr (above 0.45% by mass) can lead to the formation of coarse, brittle phases such as w-AlCuSc and large $Al_3(Sc, Zr)$ particles [110, 111]. The incorporation 0.17 wt.% Zr and 0.38 wt.% Sc to Al-7.1Zn-2.2Mg-1.6Cu-0.3Mn (SSA038) gives a room temperature yield strength of ~696 MPa with 14% elongation, and 115 MPa with 20% elongation at 300°C (Figure 2.24a). After 100 hours at 205°C, the yield strength drops to ~190 MPa (Figure 2.24b). Figure 2.24(c-e) shows the T6-tempered Al-7.2Zn-2.2Mg-1.6Cu-0.3Mn-0.18Zr-0.18Sc (SSA018) alloy with a fine, stable η′ precipitate distribution, and its microstructure resists coarsening even after prolonged thermal exposure. Beyond grain refinement, these elements also increase resistance to recrystallization, stabilize the deformed microstructure during subsequent heat treatments, and further strengthen the alloy through the formation of fine, coherent $Al_3(Sc, Zr)$ dispersoids. These dispersoids are both thermodynamically and kinetically stable during conventional heat treatment [104].

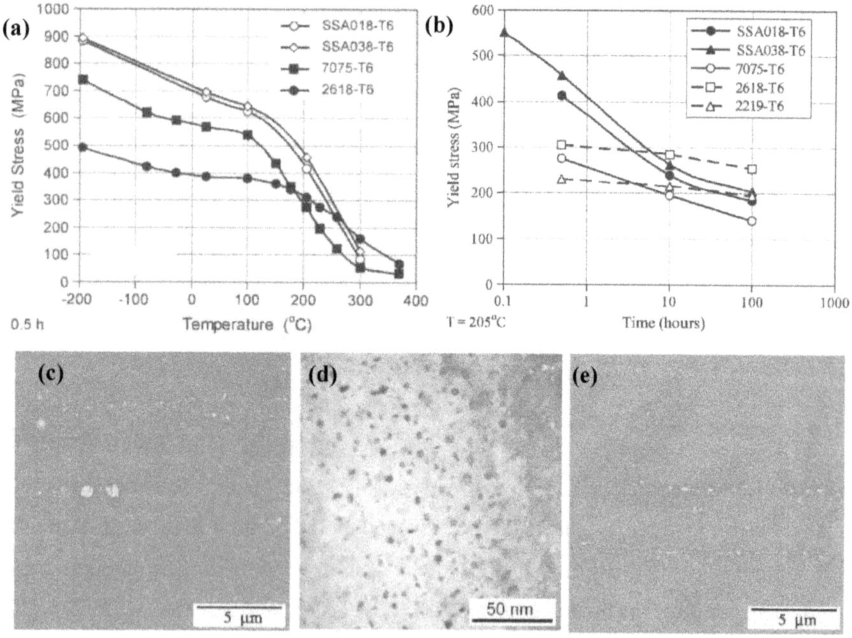

Figure 2.24 (a) Yield strength as a function of temperature for T6-tempered 7075, 2618, SSA018 (Al-7.2Zn-2.2Mg-1.6Cu-0.3Mn-0.18Zr-0.18Sc), and SSA038 (Al-7.1Zn-2.2Mg-1.6Cu-0.3Mn-0.17Zr-0.38Sc) alloys (b) Yield strength at 205°C for these alloys, plotted against holding time at 205 °C before testing. (c) Backscatter electron micrograph of SSA018 after T6 treatment. (d) TEM image highlighting η′ precipitates. (e) Backscatter electron image of SSA018 following T6 temper and 100 h exposure at 205 °C [104].

Ti also plays a crucial role, particularly in reducing hot tearing sensitivity (HTS). For example, adding 0.06% Ti to Al-6Zn-2Mg-2Cu alloys significantly lowers the HTS factor, but higher Ti content reverses this benefit [112]. Additionally, Ti refines grain size, reduces phase agglomeration, and modifies the $L1_2$ dispersoid from $(Al, Zn)_3Zr$ to $(Al, Zn)_3(Zr, Ti)$, decreasing the lattice constant and further refining precipitate size [113]. Ti also promotes the formation of $Al_{18}Mg_3Ti_2$ compounds, which serve as nucleation sites for strengthening phases during solidification and solution treatment [114, 115]. Additionally, minor Ag (0.16 wt.%) additions promote the formation and stability of GP zones and transition phases, delays over-aging, and further improves the thermal stability of the alloy [116]. Furthermore, the careful control of these microalloying strategies enables significant improvements in the thermal stability, strength, and casting performance of Al-Zn-Mg-Cu alloys. These compositional modifications of Al-Zn-Mg-Cu based alloys and their effects on room and elevated temperature mechanical properties are summarized in Tables 2.4a and 2.4b.

Table 2.4a Composition of recently developed Al-Zn-Mg-Cu based alloys in wt.%.

Alloy	Alloying elements										
	Zn	Mg	Cu	Zr	Fe + Si	Ni	Mn	Sc	Cr	Ti	Al	Ref.
7075	5.6	2.5	1.6	<0.05	<0.15	-	0.3	-	0.23	0.2	Bal.	[104]
AZMCMZS	7.17	2.2	1.58	0.18	<0.15	-	0.3	0.18	<0.05	<0.05	Bal.	[104]
AZMCMZS	7.11	2.14	1.56	0.17	<0.15	-	0.25	0.38	<0.05	<0.05	Bal.	[104]
AZMCZ	4.2	2.8	1.0	<0.01	<0.15	-	-	-	-	-	Bal.	[108]
AZMCZ	6.0	2.8	1.0	<0.01	<0.15	-	-	-	-	-	Bal.	[108]
AZMCZ	8.0	2.8	1.0	<0.01	<0.15	-	-	-	-	-	Bal.	[108]
AZMCZ	8.0	1.8	1.0	<0.01	<0.15	-	-	-	-	-	Bal.	[108]
AZMCZ	8.0	0.8	1.0	<0.01	<0.15	-	-	-	-	-	Bal.	[108]
AMZ	8.5	4.5	-	-	0.02	-	-	-	-	-	Bal.	[109]
AMZCN	7.78	4.17	4.4	-	-	3.99	-	-	-	-	Bal.	[87]

Table 2.4b Ambient and elevated temperature mechanical properties of Al-Zn-Mg-Cu based alloys.

Alloy	Heat treatment	Room-Temp Tensile Strength (MPa)	Room-Temp Yield Strength (MPa)	Room-Temp Elongation (%)	High Temp (°C)	High Temp Tensile Strength (MPa)	High Temp Yield Strength (MPa)	High-Temp Elongation (%)	Ref.
7075	T6	~620	~570	~10	300	~70	~55	~60	[104]
AZMCMZS	T6	~723	~676	~15	300	~86	~72	>20	[104]
AZMCMZS	T6	~736	~696	~14	300	~118	~115	>20	[104]
AZMCZ	T6	~462	~378	~10.8	Exposure at 300°C/500 h	~209	~75	~17.9	[108]
AZMCZ	T6	~476	~398	~10.4		~209	~71	~19.6	[108]
AZMCZ	T6	~466	~389	~11.3		~193	~71	~13.4	[108]
AZMCZ	T6	~391	~313	~11.2		~189	~77	~18.1	[108]
AZMCZ	T6	~312	~255	~15.3		~184	~67	~22.7	[108]
AMZ	T6	~503	~463	>2	300	~64	~56	>2	[109]
AMZCN	T6	~497	~414	~3.5	300	~102	~97	~15	[87]

2.10.3 Thermal stability of Al-Li alloys

Thermal stability in alloys depends on understanding property evolution rates, governed by phase transformation kinetics and dominated by diffusion processes. Exposure to room temperature in select aluminum alloys can induce natural aging, characterized by minor solute atom rearrangements in the matrix that influence overall material performance, including strength and corrosion resistance [117]. In the case of Al-Li alloys, aging triggers the decomposition of the lithium-rich solid solution and the formation of Al_3Li precipitates [118], which may eventually result in embrittlement during long-term service [119, 120]. Figure 2.25a shows the Al-Li phase diagram of different phases form in Al-Li alloys at different temperatures and Li concentrations. At low Li content, the α-Al and δ' (Al_3Li) phases forms. Upon cooling below the solvus line, Al_3Li precipitates comes out from the α-Al matrix, enhancing the alloy's strength but possibly compromising ductility if precipitation is excessive, particularly at grain boundaries [121].

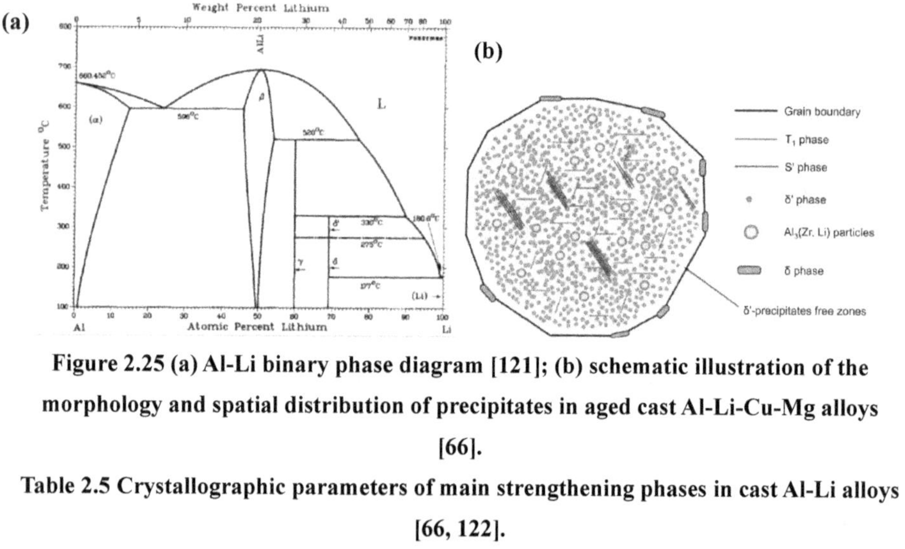

Figure 2.25 (a) Al-Li binary phase diagram [121]; (b) schematic illustration of the morphology and spatial distribution of precipitates in aged cast Al-Li-Cu-Mg alloys [66].

Table 2.5 Crystallographic parameters of main strengthening phases in cast Al-Li alloys [66, 122].

Phase	Crystal System	Lattice Parameters (Å)			Orientation	Precipitation Temperature (°C)
		a	b	c		
δ' (Al_3Li)	Cubic ($L1_2$)	4.047	4.047	4.047	Cube-on-cube	100-200 (metastable)
T_1 (Al_2CuLi)	Hexagonal	4.965	4.965	9.345	$(0001)_{T_1}//(111)_{\alpha\text{-}Al}$, $(1010)_{T_1}//(110)_{\alpha\text{-}Al}$	150-250

The addition of alloying and micro-alloying elements such as Cu, Mg, Zr, and Sc in cast Al-Li alloys leads to the formation of various strengthening phases, including δ′-Al_3Li, θ′-Al_2Cu, T_1-Al_2CuLi, and S′-Al_2CuMg, as well as dispersed phases like Al_3Zr, Al_3Sc, and core-shell structures, as illustrated in Figure 2.25b. The crystallographic details of these phases are provided in Table 2.5 [66, 122].

The first and second generations of Al-Li alloys, developed in the 1950s-60s (first generation) and 1980s (second generation), often exhibited drawbacks such as limited ductility, poor fracture toughness, and unreliable resistance to fatigue and corrosion [123]. For example, the 8090 alloy, developed by the European Aluminium Association, has been utilized in various aerospace applications, including the stringers, fuselage structure and frame of the AgustaWestland EH101 helicopter (Figure 2.26a), the wing skin of the US Fe15 aircraft, and the fuselage structures and stringers of both the Mirage fighter and the European experimental fighter FEA2000, as well as the Titan rocket (Figure 2.26b). China has also effectively used Al-Li alloys in the resource module and solar panel wings of the Tiangong-1 spacecraft (Figure 2.26c), and in the C919 passenger aircraft (Figure 2.26d and 2.26e) [8, 124-126]. However, the second-generation alloys like AA2090 (Al-Li-Cu), AA8090, and 2091 (Al-Li-Cu-Mg) experienced increased strength but reduced ductility and toughness after aging at ~70°C, which could result in embrittlement. In AA8090 alloy, this embrittlement is associated with the formation of AlLi phases at grain boundaries, Li segregation and the formation of G-P zones [127].

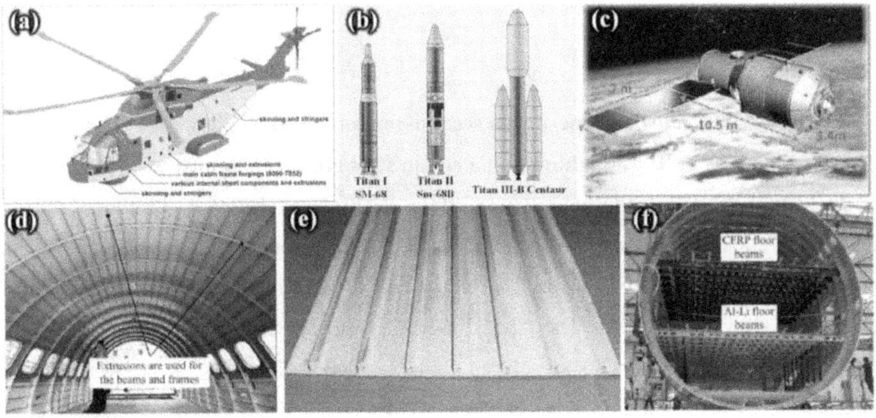

Figure 2.26 Applications of Al-Li alloys in various aerospace structures; (a) AgustaWestland EH101 helicopter; (b) Titan rocket; (c) Tiangong-1 resource module and solar panel; (d) 2099 and 2196 alloy extrusions for C919 fuselage beams and

frames; (e) 2090 alloy extruded stiffened panels for the space shuttle's external fuel tanks; (f) floor beam of the A380 [8].

The third-generation Al-Li alloys, introduced in the 1990s with lower Li content and enhanced alloying strategies, became highly suitable for modern aerospace applications. The Airbus A380 incorporates AA2099-T83 and AA2196-T8511 alloys in floor beams (Figure 2.26f), AA2196-T8511 in fuselage stringers, and AA2050-T84 for lower wing reinforcement [128]. The AA2099 Al-Cu-Li T83 alloy demonstrated good thermal stability between 200°C and 305°C, outperforming other high-temperature Al alloys in both hardness and tensile strength, while also offering reduced weight [20]. Table 2.6a and 2.6b summarizes some of the recently developed Al-Li alloys designed for high-temperature applications. Studies have shown that optimizing the formation of T1 (Al_2CuLi) precipitates is crucial for high-temperature strength. Additionally, the inclusion of 0.29 wt.% cerium (Ce) was found to inhibit the coarsening of Al_2CuLi and refine the Ce-containing Al_8Cu_4Ce phase, further enhancing thermal stability and deformation uniformity at 170-270°C [129].

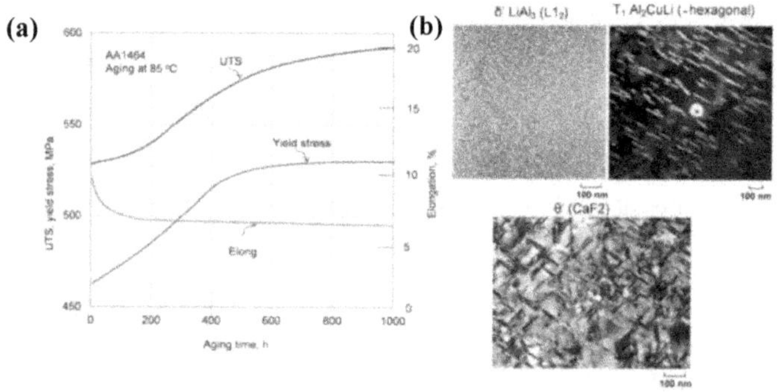

Figure 2.27 (a) Tensile properties in second-generation Al-Li alloy AA1464 during prolonged aging at 85°C, highlighting a rise in strength accompanied by a decrease in ductility [120]; (b) strengthening precipitates in third-generation Al-Li alloys [123].

Figure 2.27a and 2.27b shows the changes in mechanical properties of AA1464 alloy after prolonged exposure at 85°C, as well as the nature of strengthening precipitates in third-generation alloys [120, 123]. The Al-Li-Cu-Zr alloy C458-T861 (Al-1.8Li-2.7Cu-0.3Mg-0.08Zr-0.3Mn-0.6Zn, in wt.%), developed for aerospace applications, demonstrated stable mechanical properties after testing at 83°C, 135°C, and 177°C for up to 1000 hours, maintaining good thermal stability up to 135°C. However, exposure to higher temperatures and longer durations resulted in decreased fracture toughness [130].

Table 2.6a Composition of recently developed Al-Li based alloys in wt.%.

Alloy	Cu	Li	Mg	Mn	Zn	Si	Fe	Ni	Zr	Ti	Sc	Ce	Ag	Al	Ref.
8090	1.4	2.5	1.2	-	-	-	-	-	0.11	-	-	-	-	Bal.	[131]
2060	3.5	1.0	0.8	0.4	0.3	-	-	-	0.2	-	-	-	0.2	Bal.	[132]
ALCNM	1.22	2.42	1.42	-	-	≤0.15	≤0.15	-	0.14	-	-	-	-	Bal.	[133]
ALCNM	1.11	2.51	1.38	-	-	≤0.15	≤0.15	0.2	0.21	-	-	-	-	Bal.	[133]
ALCNM	0.96	2.33	1.22	-	-	≤0.15	≤0.15	0.4	0.19	-	-	-	-	Bal.	[133]
ALCNM	1.03	2.62	1.53	-	-	≤0.15	≤0.15	0.9	0.18	-	-	-	-	Bal.	[133]
ALCMZS	2.11	2.17	0.51	-	-	-	-	-	0.14	0.05	0.2	-	-	Bal.	[134]
ALCAMZ	4.62	1.29	0.39	-	-	-	-	-	0.14	-	-	-	0.41	Bal.	[129]
ALCAMZ	4.63	1.31	0.40	-	-	-	-	-	0.13	-	-	0.2	0.38	Bal.	[129]

Table 2.6b Ambient and elevated temperature mechanical properties of Al-Li based alloys.

Alloy	Heat treatment	Room-Temp Tensile Strength (MPa)	Room-Temp Yield Strength (MPa)	Room-Temp Elongation (%)	High Temp (°C)	High Temp Tensile Strength (MPa)	High Temp Yield Strength (MPa)	High-Temp Elongation (%)	Ref.
8090	T81	~480	~410	~12	200	~140	~128	~5	[131]
2060	T6	~540	~480	~16	300	~175	~168	~26	[132]
ALCNM	T4	~380	~324	~5	300	~83	~81	~17	[133]
ALCNM	T4	~405	~310	~7	300	~88	~86	~23	[133]
ALCNM	T4	~375	~286	~6	300	~79	~71	~58	[133]
ALCNM	T4	~354	~258	~14	300	~148	~102	~16	[133]
ALCMZS	T6	~557	~424	~7.11	225	~255	~255	~9.8	[134]
ALCAMZ	T8	~618	~604	~7	300	~246	~195	~13	[129]
ALCAMZ	T8	~625	~614	~7.5	300	~260	~206	~18	[129]

2.11 Microstructure tuning strategies to enhance heat resistance in cast Al alloys

2.11.1 $L1_2$ core-shell dispersoids

In the context of precipitation-strengthened Al alloys, traditional approaches often depends on primary alloying elements like Cu and Zn due to their high solid solubility in Al, aiming to maximize precipitation strengthening [103, 134-137]. However, the rapid diffusion rates of these highly soluble elements often lead to poor thermal stability of the strengthening precipitates, limiting their application to temperatures below 200°C.

A significant advancement in heat-resistant Al alloys emerged from Seidman et al.'s work at Northwestern University, where they developed a high-strength alloy by adding trace scandium to form coherent Al_3Sc precipitates [39]. The Sc additions to Al alloys primarily served to eliminate as-cast dendritic structures [138], inhibit recrystallization [139-141], and refine grain size [142], thereby enhancing strength and plasticity. However, Seidman et al. focused their research toward understanding the precipitation behavior of Sc. The study showed that high densities of coherent, spherical $L1_2$-structured Al_3Sc precipitates can form via conventional heat treatment: high-temperature solutionizing followed by low-temperature aging. A key advantage of Sc is its low solid solubility in Al and its lower diffusion rate relative to Cu and Zn. As a result, Al_3Sc precipitates exhibit excellent resistance to coarsening and dissolution at elevated temperatures. Even after extended exposure at 300°C, these precipitates remain stable and continue to effectively hinder dislocation motion, providing good creep resistance [37, 143].

Transition metals like Zr, Er, and Ti can form $L1_2$-Al_3TE precipitates in Al, allowing their partial replacement of Sc in $L1_2$-structured $Al_3(Sc_{1-x},TE_x)$ phases [12, 144, 145]. Notably, the addition of Zr, which diffuses more slowly than Sc, leads to the development of a stable "core-shell" $Al_3(Sc_{1-x}, Zr_x)$ phase. This structure is a result of the substantial difference in diffusion rates between Sc and Zr, which further improves the alloy's thermal stability [146, 147]. Seidman et al. also observed that minor additions of Si to Al-Sc-Zr alloys promote the uniform distribution of $Al_3(Sc_{1-x}, Zr_x)$ precipitates, resulting in finer precipitates, increased number density, and substantial improvements in mechanical properties at both room and elevated temperatures [148-150]. According to DFT calculations, Si atoms readily interact with Sc, and the diffusion of the Si-Sc complex effectively lowers the diffusion barrier for Sc in the Al matrix, thereby increasing the diffusion rate of Sc.

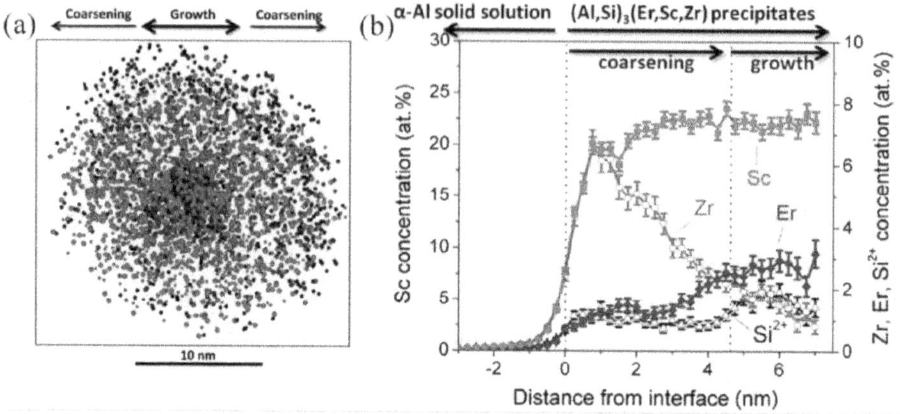

Figure 2.28 (a) APT reconstruction of a precipitate in Al-0.055Sc-0.005Er-0.02Zr-0.05Si after isothermal exposure at 400°C for 66 days; (b) corresponding concentration profiles across the matrix/precipitate boundary [151].

Expanding on these findings, researchers have engineered high-temperature resistant Al alloys by leveraging the differential diffusion rates of Sc, Zr, and Er, which results in the formation of composite precipitates containing Al, Sc, Zr, Er, and Si [152-157], as illustrated in Figure 2.28 [151]. In these precipitates, Er has the fastest diffusion rate and enhances nucleation density in the core, while Zr, with the slowest diffusion rate, improves thermal stability in the shell. Si, together with Sc, is distributed in the intermediate region to maintain an adequate volume fraction. This multi-element synergy delivers outstanding creep resistance at 350 °C. This discovery offers a key strategy for designing thermally stable Al alloys and reveals atomic-scale spinodal decomposition alongside selective recombination during the formation of "core-shell" structures. The identification of these mechanisms is significant for the advancement of solid-phase transformation theory [158, 159] and will have a lasting influence on the design and control of precipitates.

Despite the exceptional thermal stability of Al_3TE nano-precipitates, a major hurdle persists: the limited solubility (≤ 0.3%) of rare earth and transition elements in the Al matrix. This constraint impedes the formation of a sufficient volume fraction of Al_3TE precipitates, leading to a room-temperature strength considerably lower than that of conventional high-strength Al alloys.

2.11.2 Co-precipitation

In addition to facilitating solute segregation at the θ'-Al_2Cu/matrix interface, trace additions of Sc, Zr, and Ti to Al alloys can also lead to the formation of thermally stable $L1_2$-Al_3TE (TE =

Sc, Zr, Ti) precipitates within the matrix [160, 161]. For example, in Al-Cu-Sc alloys subjected to creep at 350 °C for 50 hours, Sc atoms initially segregate as a layer at the broad θ′-Al_2Cu/Al interfaces, but over time, these atoms locally transform into Al_3Sc precipitates [162]. During the growth of Al_3Sc at the interface, Sc is progressively consumed from adjacent areas. As a result, θ′-Al_2Cu precipitates in these Sc-depleted zones begin to dissolve, leading to the release of Cu into the matrix. This process leads to "precipitate splitting." Consequently, the formation of Al_3Sc destabilizes the Sc segregation layer at temperatures above 350°C, which can significantly reduce the high-temperature mechanical properties of the alloy.

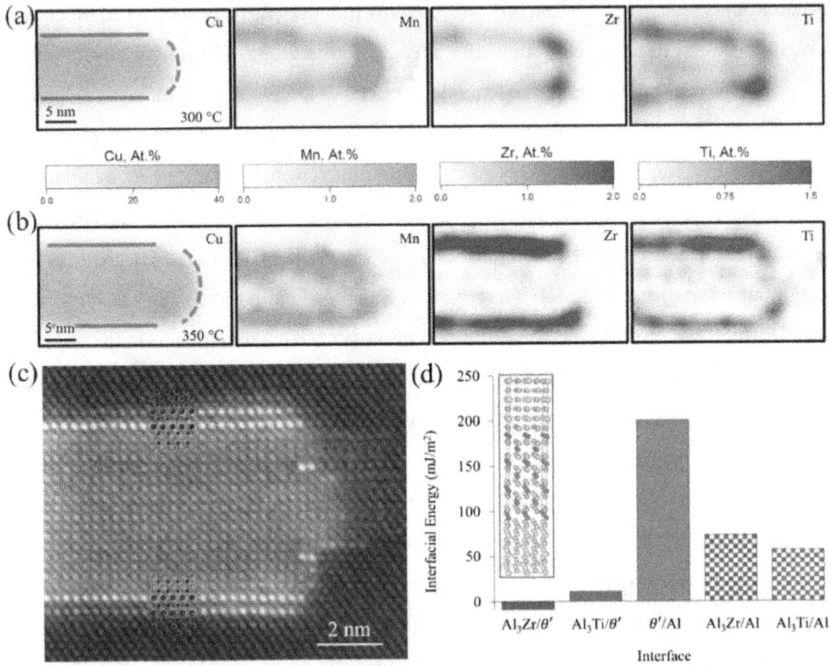

Figure 2.29 (a) and (b) 2D contour maps from atom probe tomography (APT) showing the θ′-Al_2Cu precipitates in an ACMZ alloy after exposure to 300°C for 2100 hours and 350°C for 2000 hours, respectively; (c) HAADF-STEM image of a θ′ precipitate viewed along the <110> direction of Al following exposure at 350°C; (d) DFT-calculated interfacial energies for several interfaces: Al_3Zr/θ′, Al_3Ti/θ′, θ′/Al, Al_3Zr/Al, and Al_3Ti/Al, including a representative supercell used for the Al_3Zr/θ′ interface calculation [163].

Given the excellent thermal stability of Al_3Sc and its ability to form at interfaces, a continuous distribution of Al_3TE precipitates co-precipitated at the θ′-Al_2Cu/matrix interface is expected to provide excellent resistance to coarsening at elevated temperatures compared to multi-solute segregation. Poplawsky et al. from Oak Ridge National Laboratory has shown that Mn additions can stabilize θ′-Al_2Cu precipitates long enough for the slower-diffusing Zr atoms to activate their stabilization mechanisms, thereby reducing the risk of instability associated with Sc segregation [71]. Elemental mapping at different temperatures reveals in Figure 2.29a and 2.29b that Zr and Ti exhibit similar segregation behaviors during prolonged exposure at 300°C and 350 °C. Notably, after 2000 hours at 350°C, both elements tend to segregate at the coherent interfaces of θ′, whereas after 1000 hours at 300°C, segregation is more pronounced at the semi-coherent interfaces. STEM-HAADF image show that the $L1_2$ phase begins to precipitate continuously between θ′ and the Al matrix at 350°C (Figure 2.29c), forming a co-precipitate structure with coherent interfaces. Over extended periods at high temperature, accumulated Zr and Ti at the θ′ broad face can form $Al_3(Zr_x, Ti_{1-x})$ precipitates. To better understand this mechanism, density functional theory (DFT) calculations were performed to evaluate the interfacial energies of Al_3Zr/θ′, Al_3Ti/θ′, θ′/Al, Al_3Zr/Al, and Al_3Ti/Al interfaces. The results (Figure 2.29d) indicate that the $L1_2$/θ′ interface has a lower energy than the $L1_2$/Al interface, suggesting that nucleation of the $L1_2$ phase is energetically favored at the θ′-Al_2Cu coherent interface. This preference promotes the formation of $Al_3(Zr_x, Ti_{1-x})$ precipitates at the interface, and the lower interfacial energy of the $L1_2$/θ′ interface compared to the θ′/Al interface helps to stabilize the θ′ phase through the formation of co-precipitates [163].

2.11.3 Single-solute segregation

Considering the observation that solute segregation can significantly lower interfacial energy at twin boundaries, and principles from Ostwald ripening theory, Chen et al. proposed that suppressing precipitate growth via interface segregation could enhance the high-temperature properties of Al alloys [164]. When a small amount of Sc is introduced into Al-Cu alloys, Sc atoms preferentially segregating at the θ′-Al_2Cu/Al interface during low-temperature aging, as shown in Figure 2.30. TEM analysis reveals that this segregation notably slows the coarsening of θ′ precipitates at 250°C, thereby improving their thermal stability. Additionally, the formation of Cu-Sc-vacancy complexes promotes nucleation, resulting in a narrower size distribution and finer θ′-Al_2Cu precipitates [165]. These combined effects account for the peak strength observed in alloys aged at 250°C. However, only about 7% of the total Sc segregates at the θ′-Al_2Cu/Al interface at this temperature, with the rest forming clusters in the matrix [166]. These Sc clusters are structurally undefined, have a limited effect on dislocation pinning compared to precipitates [167], and are susceptible to dissolution and rearrangement, which restricts their impact on high-temperature properties.

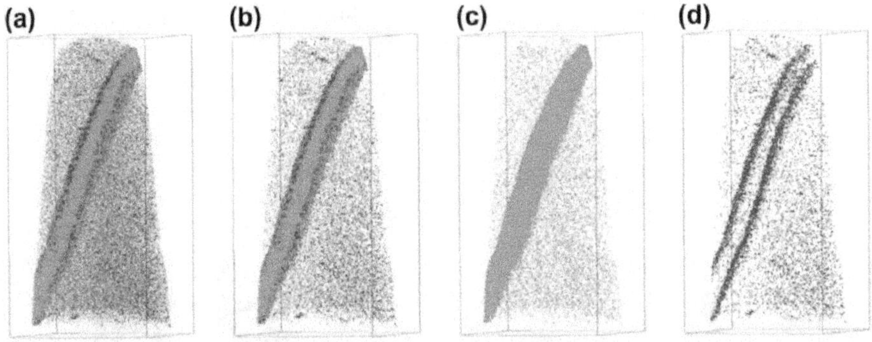

Figure 2.30 The 3DAP analysis reveals an θ′-Al_2Cu precipitate and the arrangement of Al, Cu, and Sc atoms in the Al-Cu-Sc sample after aging at 250°C for 3 hours (Al: green, Cu: orange, Sc: blue, dimensions: 80 × 80 × 180 nm); (b) highlights both Cu and Sc atoms; (c) only Cu, and (d) only Sc atoms [164].

Gao et al. optimized the heat treatment process to enhance Sc utilization in Al-Cu alloys, leading to a pronounced microalloying effect even with trace Sc additions [75]. Microstructural analysis revealed a 40% increase in θ′-Al_2Cu precipitate density and a 50% reduction in average particle size, along with a more uniform distribution. Three-dimensional atom probe tomography (APT) showed that Sc concentration at the interface was nearly ten times higher than in the matrix (Figure 2.31a). Thermodynamic modeling [168] indicated that this segregation lowers the interfacial energy of the θ′/Al interface by ~42 mJ m^{-2}, a 25% reduction from the unsegregated interface (~170 mJ m^{-2}). These microstructural improvements produce substantial mechanical property enhancements, with yield strength doubling at room temperature and creep life at 300°C/30 MPa increasing by over two orders of magnitude (Figure 2.31b). The improved thermal stability and creep resistance are attributed to the uniform precipitate distribution and Sc segregation at the interface, which reduces interfacial energy and the driving force for coarsening. In addition, Sc segregation at the θ′-Al_2Cu interface hinders Cu diffusion and growth ledge migration, further stabilizing the disc-like precipitates.

Subsequent investigations have found that elements such as Si, Ag, Zr and Ti are also capable of segregating at precipitate interfaces [71, 81, 169, 170]. To gain deeper insight, Shin et al. at Oak Ridge National Laboratory constructed models for coherent and semi-coherent θ′-Al_2Cu/Al interfaces and calculated segregation energies for 34 elements using DFT (Figure 2.32a) [171]. The results (Figure 2.32b) show that Sc segregation is energetically preferred at both interfaces, which is consistent with experimental data. Conversely, Mn and Zr are more likely to segregate at semi-coherent interfaces, where their stabilizing effect on precipitates at high temperatures is more significant due to increased interfacial energy.

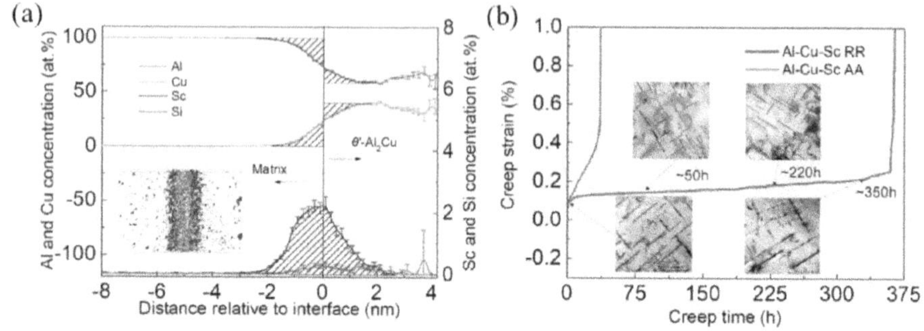

Figure 2.31 (a) Proxigram profiles of Al, Cu, Sc, and Si across Al-Cu-Sc alloys. The inset displays the corresponding APT image (20 × 20 × 30 nm³); (b) Tensile creep curves for artificially aged Al-Cu-Sc and retrogression and re-aged Al-Cu-Sc alloys, both tested at 300°C and 30 MPa [75].

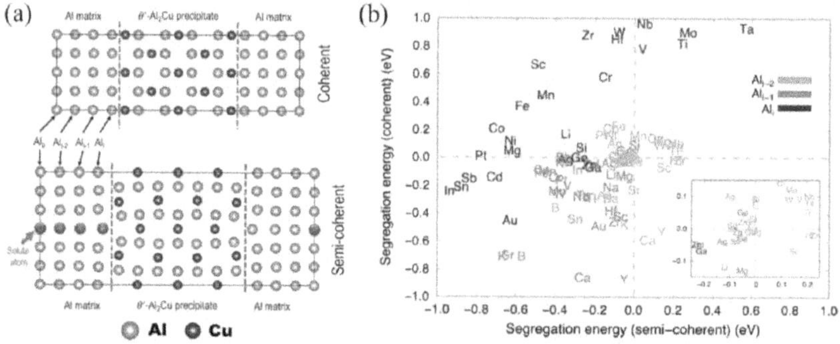

Figure 2.32 (a) Supercell models, adapted from Biswas et al., used in the current DFT calculations to represent coherent (top) and semi-coherent (bottom) interfaces between the Al matrix and θ′-Al₂Cu precipitate; (b) DFT-calculated segregation energies of solute atoms for both interfaces [171].

The underlying mechanisms stem from interfacial energy differences and solute concentration gradients, which drive solute segregation. Variations in interfacial energy create distinct adsorption sites for solute atoms near the interface, while concentration gradients intensify this segregation. Thermodynamically, solute segregation at the precipitate-matrix interface lowers interfacial free energy, thereby reducing the high-temperature coarsening rate of θ′-Al_2Cu precipitates. Additionally, chemical interactions among solute atoms at the interface further minimize this energy. These results are highly significant for understanding interfacial segregation in Al-Cu alloys and enhancing heat-resistant alloy design paradigms.

2.11.4 Multiple solute segregation

The presence of elemental segregation at interfaces can significantly influence the thermal stability of precipitates by lowering the interfacial energy. As a result, stabilizing these segregated regions can further improve the mechanical performance of alloys at elevated temperatures. In the Al-Cu-Mg-Ag-Si-Sc alloy system, the presence of various interface structures, such as Sc-rich segregation layers, C-$AlMg_4Si_3Cu$ and L phases, and the newly discovered χ-AgMg phase-partially envelops θ′-Al_2Cu precipitates. Atomic-resolution imaging and computational studies (Figure 2.33) reveal that these coexisting interface structures work together to lower interfacial energy and significantly inhibit precipitate coarsening during thermal exposure. As a result, the alloy maintains exceptional strength and heat resistance, retaining up to 97% of its yield strength during prolonged high-temperature exposure at 200°C for 100h [172].

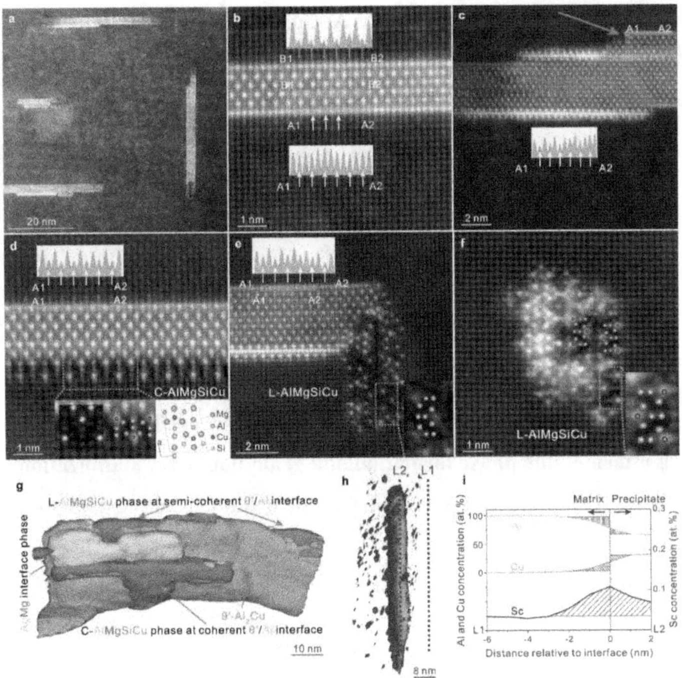

Figure 2.33 (a) Low-magnification HAADF-STEM image; (b) isolated θ′-Al_2Cu precipitate; c) AgMg interface phase present at both the upper and lower θ′/Al interfaces; (d) θ′-Al_2Cu with a C phase at the coherent interface; (e) θ′-Al_2Cu associated

with an L phase at the semi-coherent interface; (f) independently formed L phase; (g, h) multiple interface phases and segregation layers at θ′/Al interfaces identified by APT; (i) proxigram showing Sc distribution across the θ′/Al interface along L1-L2 in (h) [172].

Recent density functional theory (DFT) calculations suggest that introducing multiple alloying elements may produce a synergistic effect on interface segregation, which has become an area of active research interest [69, 173].

2.11.5 Stabilization of matrix grain boundaries by intermetallic phases

Li et al. [174] investigated the impact of Ni-containing intermetallic phases on the heat resistance of Al alloys. They identified three primary Ni-rich intergranular phases: the blocky ε-Al_3Ni, and the strip-like δ-Al_3CuNi and γ-Al_7Cu_4Ni. Among these, the Al_3CuNi and Al_7Cu_4Ni phases were found to be the most effective, providing the highest volume efficiency and the greatest contribution to high-temperature strength. For example, the presence of 1.95 vol.% δ-Al_3CuNi led to an increase of 18.89 MPa in tensile strength at 350°C. In a related study, Rakhmonov et al. [175] examined Al-Cu-Mn-Zr alloy and found that a high-volume fraction of intergranular intermetallic, such as $Al_7Cu_2(FeNi)$ and Al_9FeNi, significantly reduced the extent of θ′-precipitate-free zones (PFZ) near grain boundaries (Figure 2.34a). This microstructural refinement improved compressive creep resistance.

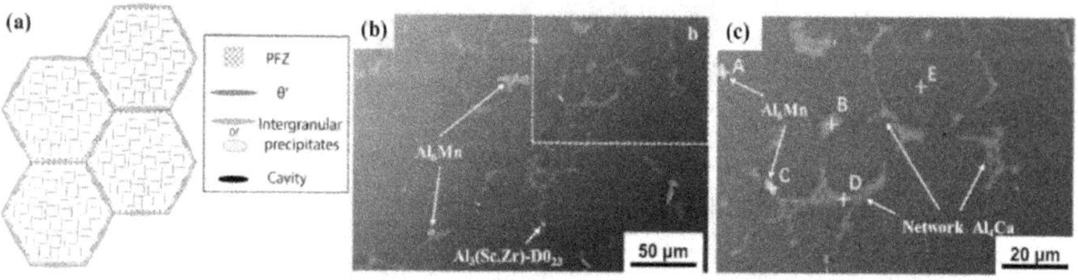

Figure 2.34 (a) Schematic representation of the microstructure in Al-Cu-Mn-Zr alloy; (b, c) SEM micrographs of Al-Mg-Ca-Sc-Zr-Mn alloy aged at 300 °C for 4 hours, illustrating intermetallic phase formation and grain boundary stabilization [175, 177].

Du et al. [176, 177] developed an Al-Mg-Ca-Sc-Zr-Mn alloy featuring a microstructure of the Al_4Ca phase forms a network along the matrix grain boundaries, while $Al_3(Sc, Zr)$ and Al_6Mn nanoparticles are finely dispersed within the grains (Figures 2.34b and 2.34c). Both the Al_4Ca network and the $Al_3(Sc, Zr)$ nanoparticles demonstrate excellent thermal stability, resulting in minimal grain growth during aging at 300-450 °C. At 300°C, this alloy achieves a tensile strength of 136 MPa and a yield strength of 114 MPa, along with notable creep resistance. The excellent high-temperature performance is primarily due to multiple strengthening mechanisms of the Al_4Ca network reinforcing grain boundaries; high-density, coherent $Al_3(Sc, Zr)$

nanoparticles; and spherical nano-sized Al_6Mn precipitates within the grains effectively pinning dislocations.

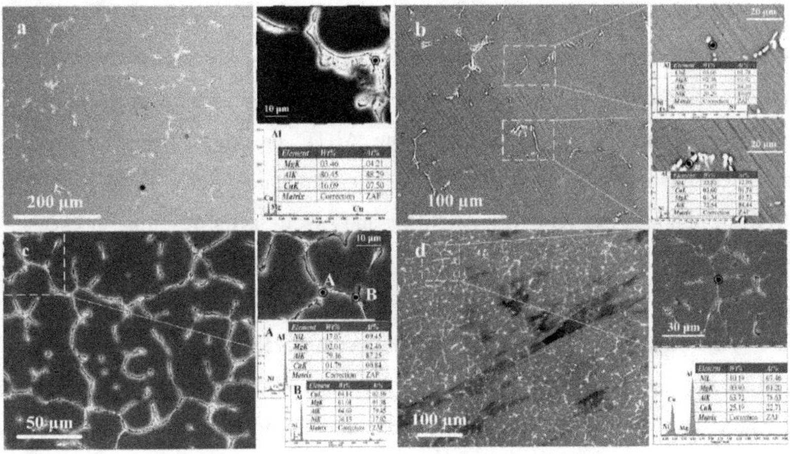

Figure 2.35 SEM-BSE images and EDS point analyses illustrate the influence of Ni addition on the microstructure of Al-2.5Li-1Cu-1.5Mg alloy: (a) base alloy; (b) 0.2% Ni; (c) 0.4% Ni and (d) 0.8% Ni [133].

Similarly, in Al-Zn-Mg-Cu alloys, the presence of Fe and Ni promotes the formation of intermetallics such as Al_9FeNi and $Al_{13}Fe_4$, which segregate at grain boundaries and effectively pin them, reducing grain growth and retaining mechanical strength at elevated temperatures [87]. In Al-Li alloys, Ni and Fe additions result in the formation of stable intermetallic like Al_3Ni and Al_3CuNi, which are distributed along grain boundaries and act as obstacles to grain boundary migration. Figure 2.35 illustrates that increasing the Ni content in the alloy promotes the formation of coarser Ni- and Cu-rich intermetallic phases at the grain boundaries. This results in more defined grain boundaries and a slight reduction in grain size [133]. Such intermetallics play a key role for grain boundary stabilization, thereby improving the alloy's resistance to grain growth and maintaining its mechanical properties under prolonged high-temperature exposures.

2.11.6 Eutectic heat-resistant Al alloys

Recent advancements in high-temperature Al alloys have focused on the development of refined eutectic microstructures, particularly in Al-Ce [178] and Al-Ni [179] based systems. Utilizing rapid solidification technology, researchers have successfully minimized the spacing between lamellar $Al_{11}Ce_3$ and needle-like Al_3Ni phases (Figure 2.36a and 2.36b), achieving interparticle spacing on the order of hundreds or even tens of nanometers. This fine dispersion increases the area density of strengthening phases and significantly enhances both room temperature and high-temperature mechanical properties. Further alloying with Mg and Si has been used to strengthen the α-Al matrix, and trace levels of transition metals such as Sc, Zr, and V facilitate the formation of $L1_2$-type Al_3TE precipitates. The use of rapid solidification helps overcome the typically low solubility of these transition metals in Al, allowing for a higher volume fraction of the strengthening Al_3TE precipitates. Recent studies have demonstrated the incorporation of complex microstructures, such as $Al_3Ni/Al_{11}Ce_3$ and Al_3Sc/Al_3Zr, in alloys like Al-Ni-Zr, Al-Ni-Sc, and Al-Ce-Si-Mg, resulting in excellent high temperature microstructural stability at 250°C [179-181]. Figure 2.37 compares stress-strain curves of Al-3.1Ni and Al-3.1Ni-0.15Zr alloys at 25°C and 250°C, showing that Zr addition increases yield strength from ~290 MPa to ~330 MPa at 25°C, and from ~110 MPa to ~200 MPa at 250°C, significantly enhancing both strength and ductility, at elevated temperature [179]. These developments underscore the promise of advanced eutectic microstructures for next-generation high-temperature Al alloys.

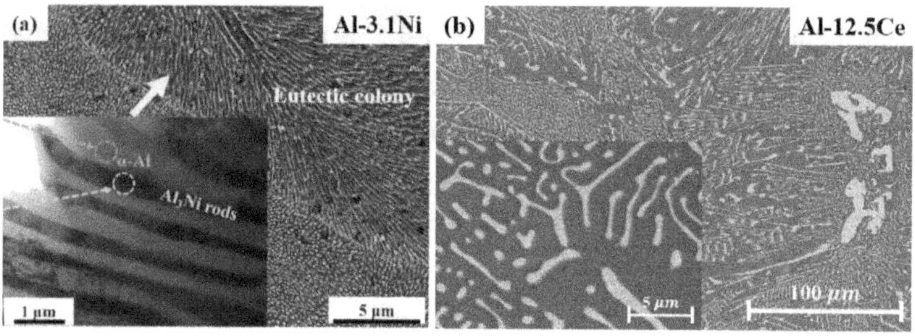

Figure.2.36 Eutectic microstructures of (c) Al-Ni [179] and (d) Al-Ce [178] alloys.

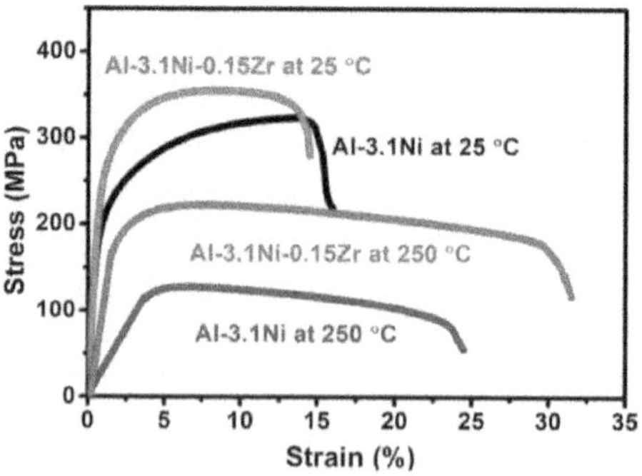

Figure 2.37 Tensile stress-strain diagrams for Al-3.1Ni and Al-3.1Ni-0.15Zr alloys (at.%) at 25 °C and 250 °C [179].

2.12 Impact of manufacturing routes on alloy thermal stability

In addition to optimizing the chemical composition, thermal stability is enhanced through manufacturing processing approaches such as rapid solidification, powder metallurgy, additive manufacturing, liquid-state engineering, and post-casting treatments. These techniques improve traditional manufacturing processes while also enabling the development of new production pathways. For instance, liquid metal processing can be incorporated as an additional step within standard casting routes. Each of these approaches influences the alloy microstructural features in distinct ways, ultimately impacting its performance under elevated temperature conditions.

2.12.1 Liquid metal engineering

Liquid metal engineering encompasses a range of physical and chemical treatments applied to molten metals with the goal of modifying their solidification behaviour [182]. By strategically combining melt chemistry with physical processing methods, this approach creates favourable conditions for nucleation and crystal growth during solidification. As a result, it contributes to improved alloy quality and homogeneity in the final microstructure.

2.12.2 Rapid solidification processing

Rapid solidification technology has been extensively investigated for aluminium alloys due to its numerous advantages over traditional ingot casting. This enhances the elevated-temperature performance of aluminium alloys by achieving high supersaturations of alloying elements within the matrix. Such supersaturation promotes the formation of fine, stable dispersoids and minimizes segregation. Among potential alloying elements, Zr, Ni, Ti, and Mn have the lowest diffusivities in Al. At 400°C they have diffusivities of 1.20×10^{-20} m^2/s for Zr, 2.05×10^{-15} m^2/s for Ni, 7.39×10^{-22} m^2/s for Ti, and 6.24×10^{-19} m^2/s for Mn [12]. Utilizing these transition elements and rare earth elements like Sc and Y, during rapid solidification can result in the formation of stable dispersoids comprising up to 30% of the volume fraction [Ref]. These microstructural features significantly enhance the high-temperature strength of Al alloys, making them viable alternatives to titanium alloys in critical aerospace components such as impellers, rib-web structures, and upper or lower wing surfaces of supersonic aircraft. This substitution can reduce overall aircraft weight by up to 15%.

2.12.3 Mechanical alloying of Al Alloys through powder metallurgy route

In the United States, Allied Signal Inc. has established large-scale production facilities for rapidly solidified aluminium alloys, followed by mechanical milling to produce alloy powders. These powders are subsequently consolidated into billets using methods such as compaction, extrusion, forging, or rolling to achieve the final product [4]. The powder metallurgy (PM) route, particularly through dispersion strengthening, has shown significant advancements for high-temperature applications and is now widely adopted by leading aluminium manufacturers. Most of these alloys are created by rapid solidification, followed by milling to obtain fine, uniform powders suitable for PM processing. Different companies have specialized in distinct alloy systems. For example, Alcoa has developed dispersion-strengthened alloys based on the Al-Fe-Ce system, Alcan has focused on Al-Cr-Mn-Zn compositions, and Pechiney has independently introduced Al-Fe-Mo-Zr alloys. Allied Signal has also expressed strong confidence in the Al-Fe-V-Si system for high-temperature performance. These alloy systems are considered promising candidates for replacing titanium-based materials in various applications [4]. Figure 2.38 shows the tensile yield strength of these PM alloys drops to ~100 MPa after thermal exposure in the range of 100-400 °C, which is notably lower than conventional 2014-T6 alloy [4].

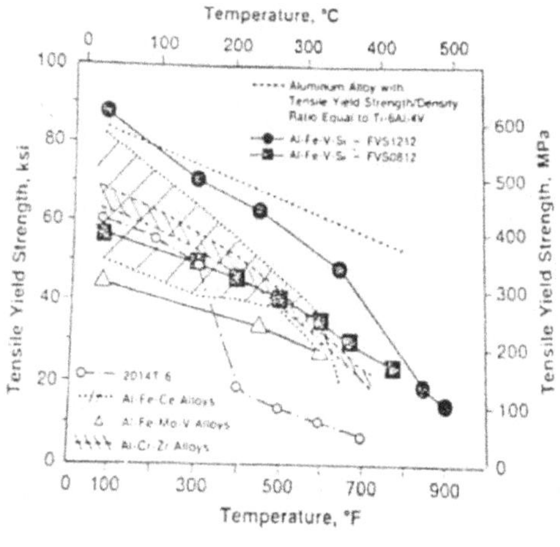

Figure 2.38 Comparison of tensile yield strengths at elevated temperatures between high-temperature aluminium alloys and conventional aerospace alloys [4].

2.12.4 Additive manufacturing

Additive manufacturing (AM), also known as 3D printing, is a cutting-edge process that fabricates components directly from digital 3D models using metallic powders or wires as feedstock. Under precise control from software and numerical systems, energy sources like lasers, electron beams, arcs, or plasma beams melt and deposit material layer by layer. This process facilitates the production of high-performance structures with complex geometries and requiring minimal post-processing [183, 184]. Among the various AM techniques for metals, powder bed fusion (PBF) and direct energy deposition (DED) are the most widely used, as illustrated in Figure 2.39a and 2.39b. PBF works by spreading a thin layer of metal powder onto a build platform and selectively melting or sintering regions with a laser or electron beam. This category includes methods like direct metal laser sintering (DMLS), selective laser melting (SLM), and electron beam melting (EBM), each offering different melting mechanisms and build rates. On the other hand, DED involves feeding powder or wire directly into a melt pool created by a focused energy source, with techniques such as laser engineered net shaping (LENS) and wire arc additive manufacturing (WAAM) included in this category [185, 186].

AM of Al alloys remains a vibrant area of research due to its economic and technological potential. Currently, Al-Si alloys such as AlSi10Mg and AlSi12 alloys are the most commonly used in AM due to their good flowability and processability [187-189]. Figure 2.39c showcases a satellite antenna bracket produced from AlSi10Mg using AM. However, due to the inherent characteristics of cast Al-Si alloys, even optimized AM processes struggle to achieve ultimate tensile strengths (UTS) above 400 MPa, limiting their use in high-performance, load-bearing aerospace components.

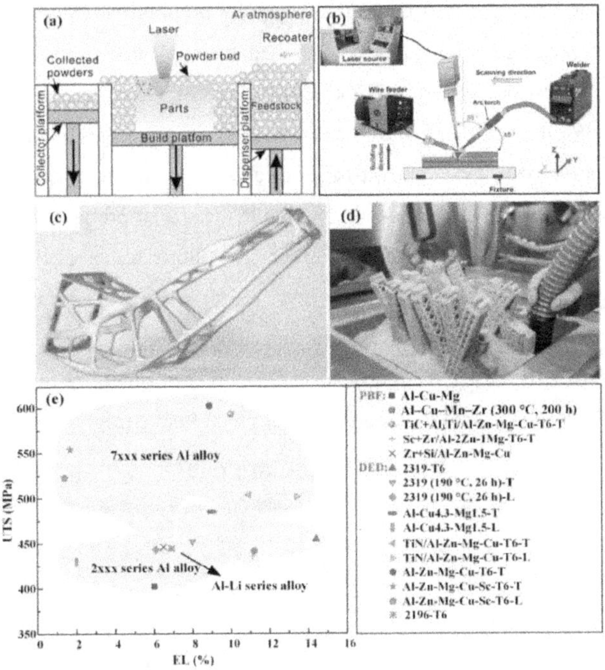

Figure 2.39 Schematic representations of (a) the Laser Powder Bed Fusion (LPBF) process [190] and (b) Wire Arc Additive Manufacturing (WAAM); (c) AlSi10Mg satellite antenna bracket produced by additive manufacturing; (d) aircraft bulkhead fabricated

from Scalmalloy; and (e) comparison of the tensile properties of 2xxx, 7xxx, and Al-Li series alloys manufactured using additive techniques [8].

To address these limitations, significant efforts have been made to develop high-strength Al alloys specifically for AM. Notably, Airbus introduced Scalmalloy, a high-strength Al alloy with a room-temperature tensile strength of 520 MPa, which has been successfully used in structural parts for the A320 aircraft (Figure 2.39d). Similarly, the 7A77.60L alloy developed by Hughes Research Laboratory (HRL) offers forged-equivalent strength above 600 MPa and is now being applied by NASA to large-scale aerospace components. Recent advancements in China have also led to the development of a new high-strength AM Al alloy with a tensile strength ~560 MPa, surpassing the performance of Scalmalloy and meeting the demands of advanced manufacturing sectors such as rail transit and aerospace [8].

Research continues to expand in this field, with strategies like alloying, grain refinement, and the introduction of reinforcing particles being explored to overcome the challenges of AM in Al alloys. For example, the addition of TiC and Al_3Ti particles to Al-Zn-Mg-Cu alloys has been shown to promote heterogeneous nucleation, suppress grain growth, and eliminate columnar grains and hot cracking during laser powder bed fusion (LPBF), resulting in alloys with impressive mechanical properties (YS of 485 MPa, UTS of 593 MPa, and elongation of 10%) [190]. Figure 2.39e shows a comparative data on the tensile properties of various AM-prepared Al alloys, including 2xxx, 7xxx, and Al-Li series alloys. In addition, laser powder bed fusion (LPBF) processing leads to a significant reduction in the size of primary Al_3Sc particles in Al-Sc alloys compared to conventional casting. Figure 2.40 illustrates the evident effect of cooling rate on Al_3Sc particles. In cast samples (Figure 2.40(a-c)), particles larger than 1 μm are commonly observed, whereas in LPB-fabricated alloys (Figure 2.40d), the particle size is refined to ~30 nm [191-193]. This significant refinement is attributed to the rapid cooling rates inherent to LPBF, which promote the formation of much finer particles during solidification.

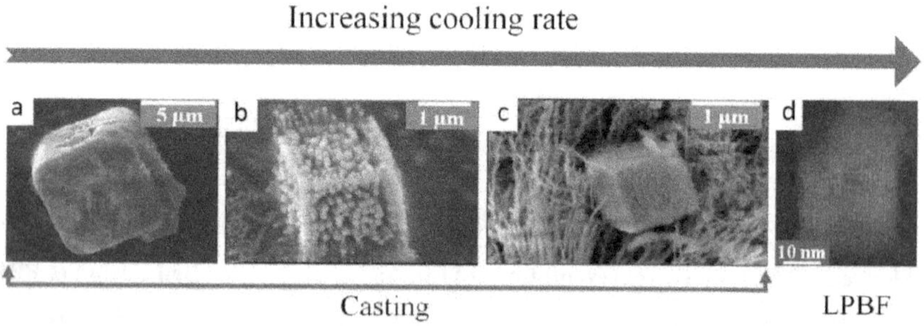

Figure 2.40 SEM images show cube-shaped primary Al_3Sc particles formed during casting at different cooling rates: (a) 1 Ks^{-1}, (b) 100 Ks^{-1}, and (c) 1000 Ks^{-1} [191]. (d) STEM image of an Al-Mn-Sc-Zr alloy reveals a primary Al_3Sc particle produced by LPBF at an cooling rate of 10^6 Ks^{-1} [192].

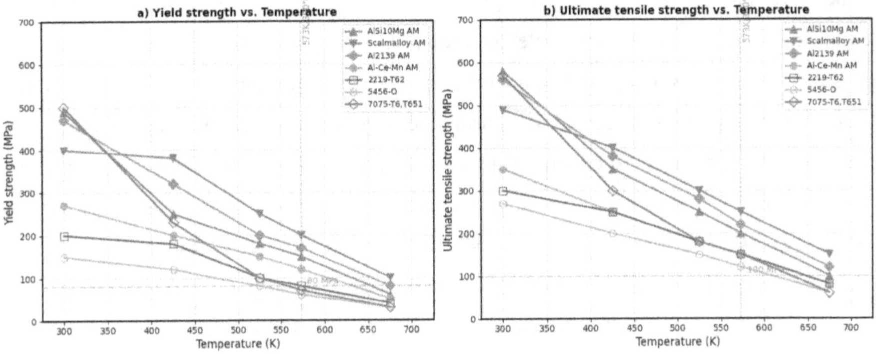

Figure 2.41 Tensile properties of various Al-alloys produced by PBF-LB additive manufacturing (AM) and conventional processing at different test temperature [194-196, 14].

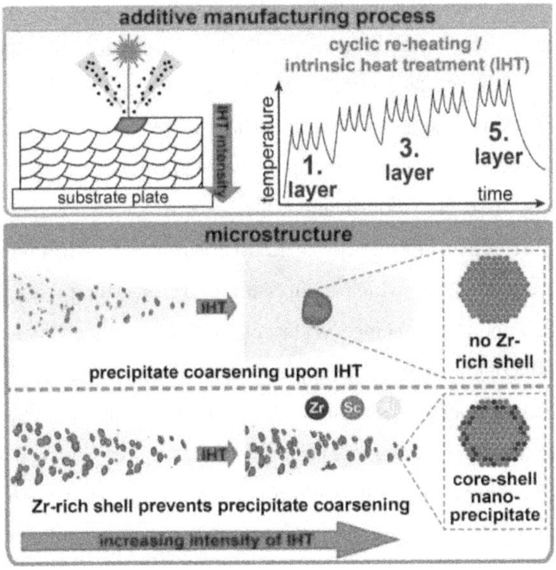

Figure 2.42 Impact of core-shell dispersoids on thermal stability in additively manufactured Al-Sc-Zr alloy [197].

Building on these advances, recent efforts have also focused on enhancing the high-temperature mechanical performance of AM Al alloys. Figure 2.41 illustrates conventional and AM Al alloys, such as AlSi10Mg and 7075-T6-exhibit a significant drop in both yield and ultimate tensile strengths as temperature increases, often retaining less than 15% of their room temperature strength at 300°C. Although some AM alloys like Al-Ce-Mn and Scalmalloy demonstrate better performance at elevated temperatures, they are often limited by relatively low strength and ductility at room temperature [194-196]. One challenge with this approach is the tendency for precipitate coarsening during the in-process heat exposure, which diminishes their strengthening effect. To counteract this, modified solidification strategies were employed to promote the formation of a Zr-enriched shell around the Al_3Sc core, enhancing precipitate stability. Figure 2.42 shows the Zr-rich shell effectively inhibits coarsening, retaining the mechanical benefits of the fine precipitates [197]. This strategy can be extended to other precipitation-strengthened alloys processed via laser-based additive manufacturing.

2.12.5 Surface re-melting process

Recently, extensive studies have explored the alternative approach of remelting the surface and sub-surface by LASER or Arc to control the material characteristics of an alloy [198, 199]. The LASER or Arc acts as the concentrated heat source, penetrating a certain depth in the substrate to melt and solidify the cast microstructure into a different microstructure to enhance mechanical behavior. Like surface remelting, surface alloying by LASER or Arc is another alternative to induce microelements on the surface to regulate the precipitation behavior, microstructure and mechanical characteristics of the Al alloys [199, 200]. Among the different energy sources used for surface modification, the LASER source has received the maximum attention from the researchers [199-202]. The LASER surface modification technique has been subdivided into five categories, which comprise (i) laser surface melting, (ii) laser surface alloying, (iii) laser composite surfacing, (iv) laser surface cladding, and (v) laser shock peening [199, 200]. The effectiveness of these approaches is demonstrated in Figure 2.43, which presents EBSD maps of laser surface remelted Al-Cu alloys with and without the addition of Al-3Ti-1B grain refiner, across a range of scan speeds (energy densities). Without grain refiner, coarse columnar grains persist regardless of processing conditions. In contrast, the addition of Al-3Ti-1B results in a uniform, fine equiaxed grain structure, even at higher scan speeds [202]. This highlights that laser-based surface remelting treatments combined with grain refiners, can significantly enhance microstructural refinement. However, surface modification of Al alloys with LASER is challenging due to the lower melting point of aluminum and the poor absorption of electromagnetic radiation (only 7%) due to the high density of free electrons [16]. Despite this, using LASER treatments promises better grain orientation, phase distribution, refinement, phase stability and improved mechanical behavior [201, 203]; however, the cost of LASER equipment and the requirement of higher energy density for Al alloys [200] failed to find cost-economic applications in the industry, particularly in developing countries, although they are employed in Western industries for repair operations.

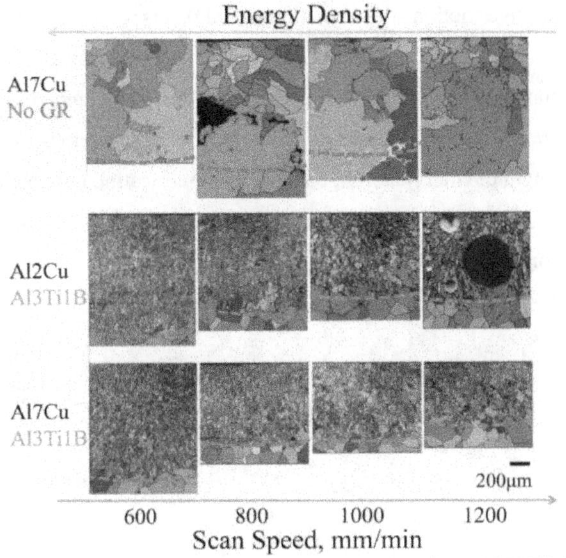

Figure 2.43 EBSD mapping of Laser Surface-Remelted Al-Cu alloys without and with Al3Ti1B grain refiners at a scan speed of 600-1299 mm/min [202].

An alternative solution to this problem is to use much more economical arc energy sources, such as tungsten inert gas (TIG) or plasma arc, for the surface modification of Al alloys. Recent research on arc surface remelting (ASR) methods has shown that they can greatly improve the microstructure and mechanical strength of materials [198, 204, 205]. However, with variations in Al alloy composition, the ASR method behaves differently. For instance, the ASR of the Al7075 alloy enhances microstructural refinement (Figure 2.44(a-c)) but reduces the intermetallic precipitates by the dissolution of phases, leading to a decrease in hardness of 20% compared to base material (Figure 2.44d) at ambient temperature [198]. On the other hand, Al-Si alloys demonstrated significant grain refinement and precipitation density after ASR, which substantially improves hardness and strength compared to the as-cast condition in an ambient environment [205]. However, the addition of alloying elements such as Cu [205] and Mn [204] on the surface of Al alloys by ASR often resulted in unfavorable precipitates including Al(Cu, Si) and Al(Mn, Si) which inhibited grain refinement and initiated lamellar cracking during solidification at the interface of the fusion zone (FZ) and the partially melted zone (PMZ), which led to the reduction in properties. On the contrary, surface alloying of Al-Cu alloys with Al_2O_3 [206] and Fe-Si [207] proved beneficial for property enhancement, including wear resistance. Al_2O_3 ceramic particles and the formation of AlFeSi ternary intermetallic compounds (IMCs) substantially improved the hardness and wear properties [206, 207]. A novel technique, additive friction stir deposition (AFSD), has recently emerged for developing heat-resistant aluminum alloys [208]. Studies have demonstrated that AFSD alloy AA9052 has exhibited approximately 22% and 49% higher yield strengths compared to forged alloy at 240°C and 340°C, respectively [208]. Nevertheless, the AFSD process is subject to limitations, such as the fixed length of filler material, challenges in fabricating intricate geometries, a lack of control mechanisms, and overall higher setup costs, rendering it unsuitable for current industrial applications.

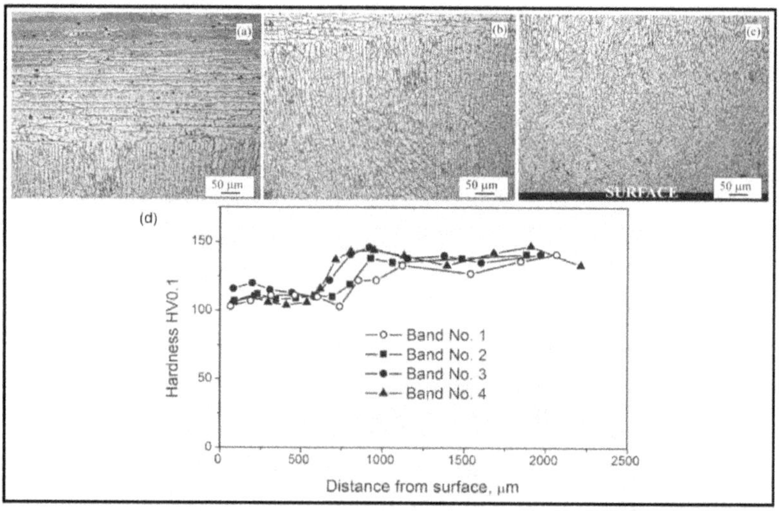

Figure 2.44 Microstructure of arc-surface re-melted Al7075 alloy (a) partially re-melted zone; (b) central part of the re-melted zone; and (c) the surface layer; and (d) variation in hardness with respect to the distance from the surface.

2.13 Progress in aluminium alloy development for aerospace applications

Williams and Starke [11] have discussed the progression of Al alloys, from their manufacturing processes to their application in aerospace components such as in engine components, turbine impellers, and airframe parts exposed to operational temperatures up to 400°C. While heat-resistant materials such as titanium or superalloys could be used to overcome these temperature limitations, these alternatives often result in increased weight and cost. Therefore, the ongoing development of Al alloys for high-temperature aerospace applications is motivated by the demand for lightweight materials that retain their structural integrity under thermal stress. The traditional 2xxx series and 7xxx series alloys degrading significantly above 150°C due to microstructural instability. To overcome these, advanced processing techniques such as rapid or sub-rapid solidification processing (RSP), mechanical alloying, powder metallurgy (PM), additive manufacturing (AM) and new strengthening approaches e.g. alloying, purification, refinement and ceramic particle reinforcement can produce high-performance materials suitable for critical parts. Al-based composites-both continuous and discontinuous fiber-reinforced are under active development. Polymer composites also contribute to specific aerospace applications. Notable examples include GLARE (glass-reinforced) and ARALL (aramid fiber-reinforced) hybrid laminates, which are being engineered for excellent fatigue and fracture resistance. Al-Li alloys, including third-generation grades like 2195 (YS ~550 MPa at room temperature, with good stability up to 200°C) and 2198, have been optimized for reduced density (10-20% lighter than traditional Al alloys) and improved stiffness. However, Al alloys face challenges due to impurities like Fe and Si, which can degrade fatigue performance. These impurities may form coarse primary phases when their solubility limits are exceeded. Modern aircraft such as the Boeing 777 and Airbus 340 incorporate these progressively improved alloys, as noted by Williams and Starke [11]. Figures 2.45 and 2.46 illustrate the gradual enhancements in yield strength and toughness in advanced Al-alloys applied in commercial airbus aircraft [8, 11]. However, despite more complex alloy systems and higher alloying content, heat-resistant Al alloys have not surpassed 100 MPa tensile strength at 350°C and use above 300°C remains difficult. Future development in cast heat-resistant Al alloys should focus on: intrinsic heat resistance of secondary phases, phase interface modification, grain boundary stabilization and diffusion-controlled phase transformation.

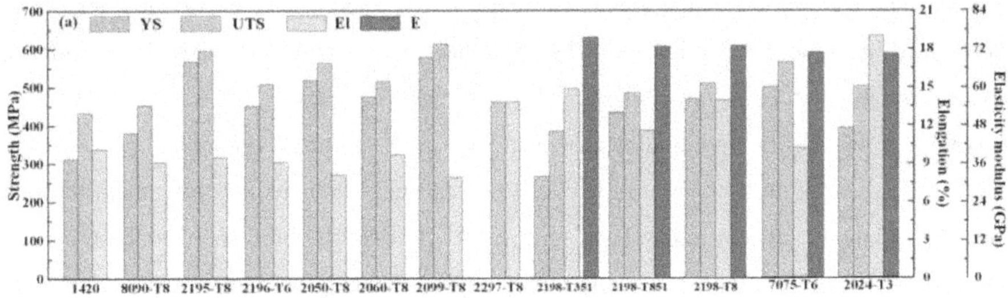

Figure 2.45 Comparison of the tensile strengths of 2XXX, 7XXX and Al-Li Al alloys [8].

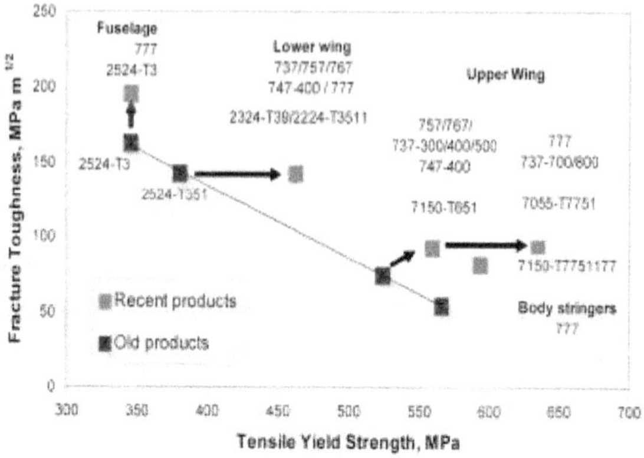

Figure 2.46 The improved balance of strength and toughness in advanced Al-alloys applied in commercial Airbus aircraft [11].

2.14 Improving performance by optimizing heat-treatment procedure

Optimizing heat treatments for 2XXX series Al-Cu-Mn alloys significantly enhance microstructure uniformity, precipitate distribution, and mechanical properties. Homogenization treatment eliminates brittle intergranular phases formed during solidification, which can impair workability and performance. For example, a two-stage homogenization process involves 470°C for 12 hours followed by 480°C for 8 hours. This dissolves low-melting eutectic phases in the first stage and completely dissolves the S-Al_2CuMg phase in the second, avoiding overburning while delivering a better homogenization effect than single-stage methods [212]. In alloys containing transition elements like Sc, Zr, Nb this two-step approach promotes uniform nucleation of fine, dense Al_3TE dispersoids, which hinder recrystallization which has increased YS by 6.4%, UTS by 6.3%, and EL by 67.6%, in 2195 alloy [213, 214]. Solution treatment requires precise temperature control due to the narrow gap between solution and eutectic melting points, which allows for full dissolution of soluble phases into a supersaturated solid solution. Double-step aging in Al-Cu-Mn alloys results in higher fracture toughness and YS with only a slight decrease in UTS, while optimizing precipitate-dislocation interactions to enhance strength, hardness, and resistance to stress corrosion cracking (SCC).

For 7XXX series Al-Zn-Mg-Cu alloys, such as 7075, heat treatment optimization focuses on balancing high strength with improved corrosion resistance and fracture toughness through homogenization, solution, and aging processes. Two-stage homogenization mitigates intergranular phases and minimizes eutectics, thereby enhancing alloy workability, similar to approaches in other series [215]. Solution treatment benefits from techniques such as two-stage solid solution, contact solution treatment (CST) [216], electropulsing treatment (EPT) [217], and salt-bath rapid heating [218], which accelerate phase dissolution; for example, CST dissolves second phases in 7075 within 5 minutes, producing finer precipitates and slightly higher post-aging strength than traditional furnace methods, while EPT similarly expedites formation of supersaturated solid solutions [216, 217]. Aging progresses from T6, which offers peak strength but low SCC resistance, to T73, T76, and T77, which coarsen and disperse η at grain boundaries for better corrosion resistance, with a 10-15% strength reduction in T7X [212, 213, 218, 219]. Advanced heat treatments such as retrogression and re-aging (RRA) combine T6 strength with T7X corrosion resistance by partially dissolving and re-precipitating GP zones and η′ within grains while coarsening of the equilibrium η phase at grain boundaries to enhance intergranular stability [214, 220, 221]. Non-isothermal aging (NIA) further optimizes performance by promoting fine intragranular precipitates for strength and discontinuous, coarser boundary precipitates with narrowed precipitate-free zones (PFZ) for SCC resistance [222].

2.15 In-situ deformation behaviour

In situ deformation behaviour observed via scanning electron microscopy (SEM) provides critical real-time insights into the micromechanical failure mechanisms of high strength cast Al alloys. Unlike ex-situ tests, which only provide after-the-fact insights, in situ SEM enables the visualization of dynamic processes such as crack initiation, propagation, and blunting at the microscale, offering a deeper understanding of the interplay between microstructural heterogeneities and mechanical response. The Al-Zn-Mg-Cu alloys are widely used in aerospace and automotive sectors due to their superior strength, low density, and fatigue resistance [8, 223]. However, they remain prone to fracture and stress corrosion cracking due to the microstructural heterogeneities such as grain boundaries, intergranular/intragranular precipitates, and second-phase particles [96, 99, 101]. Strengthening via heat treatment or microalloying with Zr, Ti, Sc, or Er refines the structure, promoting precipitation sequences from supersaturated solid solutions to GP zones, η' and η phases, enhanced by dispersoids such as core-shell $Al_3(Sc, Zr)$ or $Al_3(Zr, Ti)$ for grain refinement and dispersion strengthening [104, 224, 225]. The degradation involves incubation, nucleation at microstructural sites, and progression to void nucleation, decohesion, or particle fracture, influenced by particle volume fraction, distribution, and shape [226]. In situ SEM facilitates direct observation of these phenomena, revealing localized plastic flow and failure initiation. Several studies [227-229] have explored the effect of grain boundary phases, precipitates and dispersoids on crack growth, particle geometry via modelling, and crack initiation, propagation, and blunting in different Al series. For instance, investigations [229, 230] detailed Si particles role in damage evolution and fracture in cast Al-Si-Mg alloys. M. Besterci et al. [231] investigated that increasing Al_4C_3 carbide and Al_2O_3 oxide fractions (up to 12 vol%) led to rupture by side crack linkage perpendicular to loading. The deformation behaviour of Al-11Ce-0.4Mg alloy [228] under in-situ FESEM reveals that crack initiation begins at the centre of the U-notch (Figure 2.47a). The crack propagates through the matrix perpendicular to the loading direction (Figure 2.47b), then diverts around eutectic colonies, continuing in the matrix and occasionally branching (Figure 2.47c). The crack path (Figure 2.47d) shows a tortuous trajectory, repeatedly redirected by eutectic colonies. The load versus crosshead displacement curve (Figure 2.47e) demonstrates quasi-brittle behaviour, with crack initiation at peak load and red dots indicating the points corresponding to Figure 2.47(a-c). Recent work [104, 225, 232] has focused on ex-situ stresses for crack growth, with limited in-situ SEM tests at micrometer scales. However, in-situ studies on multiphase Al-Zn-Cu-Mg alloys remain scarce, particularly regarding Al_3TE (Sc, Zr, Ti) dispersoids and eutectic phases impact on crack initiation, propagation, and fracture in 7xxx-series alloys. This gap highlights the need for expanded in-situ SEM research to optimize microstructures and enhance mechanical properties.

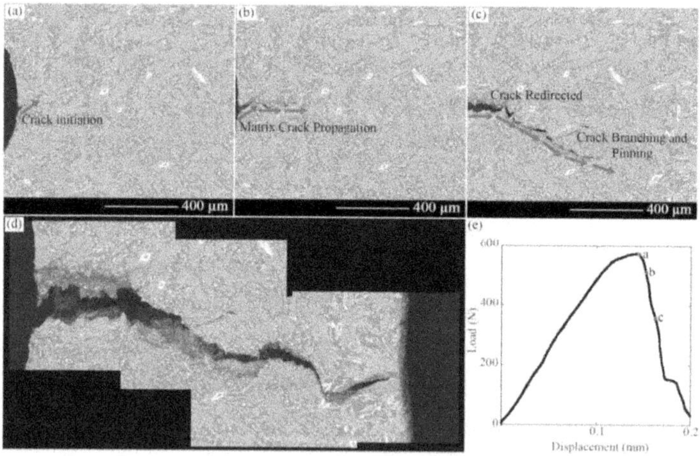

Figure 2.47 In-situ SEM tensile deformation behaviour of Al-11Ce-0.4Mg alloy (a) crack initiation at the notch tip during in-situ testing; (b) crack propagation through the alloy matrix; (c) crack redirection into the matrix, bypassing a large Al-Ce eutectic colony, with branching observed and arrested at the colony edge; (d) stitched image showing the full crack path, predominantly through the matrix; (e) load versus crosshead displacement curve for the single U-notch specimen, exhibiting quasi-brittle fracture behaviour; red dots mark positions corresponding to images (a), (b), and (c) [228].

2.16 High temperature deformation

The high-temperature deformation behavior of Al-Zn-Mg-Cu alloys is governed by dislocation-precipitate interactions, which evolve significantly after microalloying additions. In conventional 7075Al and 7085Al alloys, elevated temperatures accelerate the coarsening of metastable η' ($MgZn_2$) precipitates and their transformation into stable η phases [104, 105, 108]. This process facilitates dislocation movement via Orowan looping, which lowers precipitate shearing resistance and leads to thermal softening and reduced strain hardening. Figure 2.48(a-d) shows that as temperature increases from 100°C to 175°C, 7085Al alloy exhibits a progressive coarsening and discontinuity of grain boundary precipitates (GBPs). This is accompanied by a notable increase in the precipitate-free zone (PFZ) width, from ~42 nm to ~108 nm [105]. For 7075Al alloy, the fracture surface at 350°C (Figure 2.48(e-f)) shows local necking and numerous tiny dimples, indicating ductile fracture with serpentine sliding. At 450°C, dimples become deeper and more coalesced (Figure 2. (g-h)), with enhanced serpentine sliding and tenacity nests, reflecting increased vacancy migration and dislocation activity [105, 106]. The addition of Sc, Zr, Ni, or increasing Cu content leads to the formation of stable dispersoids like $Al_3(Sc, Zr)$, T-$Al_6Mg_{11}Zn_{11}$, and $Al_3(Cu, Ni)_2$. These dispersoids improve dislocation and grain boundary pinning, shifting deformation from Orowan looping to stronger dislocation-precipitate entanglement and shearing resistance at 300°C. This results in significantly improved creep rupture life (over 10-fold improvement), maintains yield strength through reduced coarsening rates of T' and η' phases, as stable phases effectively retard dislocation motion and grain boundary sliding [87, 104-108].

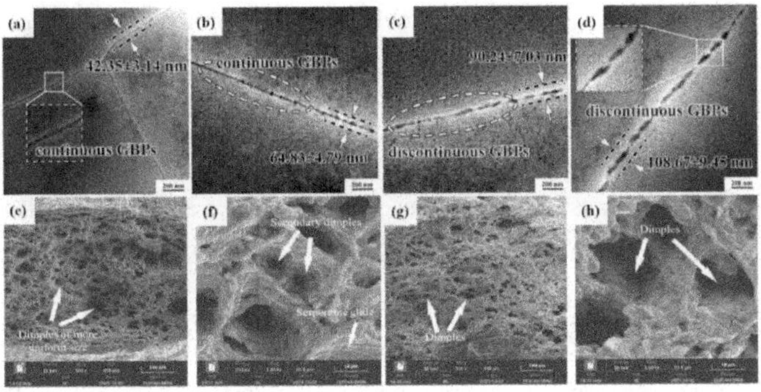

Figure 2.48 (a-d) Grain boundaries in the 7085Al alloy after tensile testing at 100°C, 125°C, 150°C and 175°C, respectively; (e-h) SEM micrographs of the fracture surfaces of 7075Al alloy tensile tested at (e, f) 350°C; (g, h) 450°C with a strain rate 0.001 s^{-1} [105, 233].

One of the prime motivations for developing Al-Cu-Mn-Zr alloys is the need for lightweight components for the aerospace sector that can withstand high temperatures [69, 234]. The high-temperature deformation resistance in Al-Cu-Mn-Ni-Zr based RR350 and Al-Cu-Sc alloys results from precipitate stabilization and grain boundary pinning, which enhance thermal stability and limit plastic deformation up to 350°C. In Al-Cu-Ni alloys, microalloying with 0.2 wt.% Mn and 0.17 wt.% Zr facilitates solute segregation to θ' precipitate-matrix interfaces. This segregation reduces interfacial energies and inhibits coarsening or transformation of θ' precipitates, as confirmed by atomic-scale analyses and first-principles calculations. As a result, it sustains yield strength increments (15-26 MPa from aging) by impeding dislocation motion and absorbing strain energy. The RR350 alloys leverage heat-resistant intermetallics (Al_3CuNi, Al_7Cu_4Ni phases) formed during solidification for grain boundary obstructions, combined with dispersoids from solution and aging treatments. This results in fish-scale fracture patterns and enhanced tensile strength at 300°C; however, excessive brittleness may cause premature cracking, often characterized by cleavage surfaces and dimples [69, 74, 235]. The T6 + 300°C/100 h re-aged Al-8.4Cu-2.3Ce-1.0Mn-0.5Ni-0.2Zr alloy [72] exhibits dense dislocation tangles around nano-sized T and θ′ precipitates (Figure 2.49(a-c)), which enhance high-temperature strength at 400°C. The re-precipitated T phase contributes significantly to yield strength, while coarsening of θ′ phases at higher temperatures reduces this effect and increases plasticity. Figure 2.49d illustrates that grain boundary intermetallics act as barriers to dislocation motion. The addition of Sc in Al-Cu alloys, such as in 2219-Sc-Zr or Al-Cu-Sc alloy systems, further enhances these mechanisms via the formation of coherent Al_3Sc or $Al_3(Sc,Zr)$ dispersoids [79, 164]. These dispersoids provide pinning against dislocation glide and grain boundary migration, promoting ductile fracture resistance at elevated temperatures. The in-situ TEM tensile test (T = 250°C, t = 3 h) of an Al-Cu-Sc alloy reveals a crack perpendicular to the tensile axis, with two orthogonal θ' precipitate variants present ahead of the crack tip (Figure 2.49e). Applying tensile stress, interfacial decohesion occurred at the precipitates ahead of the crack tip, leading to microcracks (Figure 2.49f). Figures 2.49g and 2.49h show crack propagation along certain matrix/precipitate interfaces, forming orthogonal growth segments that align with the precipitate orientations [164]. However, Sc additions reduce coarsening rates, increase creep resistance, and promote uniform deformation by stabilizing θ' precipitates, with the 2219ScZr alloy showing up to 42% higher strength retention at 250°C compared to Sc-free counterparts.

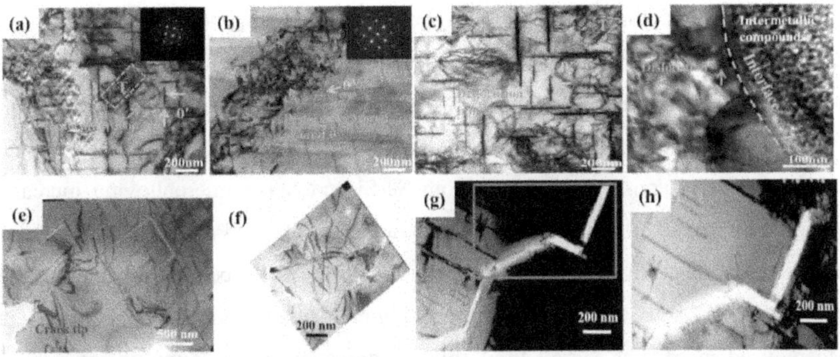

Figure 2.49 (a-c) BF-TEM images of T6 Al-8.4Cu-2.3Ce-1.0Mn-0.5Ni-0.2Zr alloy after 400°C tensile testing and re-aging at 300°C for 100 h show θ′ and T phases obstructing dislocation movement; (d) intermetallic at the grain boundary acting as a barrier. (e-h) in-situ TEM tensile tests on Al-Cu-Sc alloy aged at 250°C for 3 h reveal crack growth driven by interfacial decohesion at θ'/matrix interfaces; (e) showing crack and precipitate distribution; (f) magnifying interfacial voids; (g, h) crack propagation along matrix/precipitate interfaces [72, 164].

2.17 Creep properties

The excellent high-temperature stability of Al-Cu based alloys also enhances their creep properties. Miligan et al. [236] demonstrated that Al-Cu-Mn-Zr (ACMZ) alloys exhibit excellent creep resistance at 300°C compared to conventional Al-Cu and Al-Sc alloys, primarily due to the thermal stability of θ′ precipitates, which resist transformation to the less effective θ phase after 200 hours of preconditioning at 300°C. The segregation of Mn and Zr to θ′ interfaces enhance this stability, strengthening the α-Al grain interiors and contributing to higher yield strengths at 300°C (96-105 MPa for ACMZ vs. 56 MPa for 206). At low stresses, ACMZ alloys show diffusional Coble creep ($n \approx 1$), where deformation shifts to grain boundaries because θ′ precipitates impede dislocation glide in the grains. In contrast, the 206 alloy exhibits dislocation creep ($n \approx 4$) even at low stresses due to precipitate instability and weaker grain interiors. At higher stresses, all alloys transition to dislocation creep ($n \approx$ 4-7), but ACMZ alloys perform better. A significant factor is the impact of grain boundary (GB) intermetallic in RR350 ($F_c \approx 42\%$, containing Al, Cu, Ni, Fe, Mn, Co), which impede vacancy diffusion, effectively increasing the "effective grain size" (93 μm) and reducing steady-state creep rates by about one order of magnitude compared to Al-7Cu ($F_c \approx$ 9.8%-21%). This mechanism is described by a modified Coble creep model, in which the intermetallic limit grain boundary-controlled deformation without significantly impacting the alloy monotonic strength (RR350: YS ≈ 96 MPa; Al-7Cu: YS ≈ 105 MPa) due to their large inter-particle spacing.

Gao et al. [237] investigated co-stabilization of θ′-Al_2Cu and Al_3Sc in Al-1.08 at.% Cu with additions of 0.18 at.% Sc, 0.18 at.% Sc + 0.05 at.% Si, and 0.18 at.% Sc + 0.05 at.% Zr after aging and thermal exposure at 300°C. Both Sc-Si and Sc-Zr microalloying increased the thermal stability of θ′-Al_2Cu, but higher Si contents accelerated Al_3Sc coarsening, limiting high-temperature benefits. The Sc-Zr combination delivered the best stabilization of the dual precipitate formation at 300°C. Atom probe tomography results indicate that θ′ stabilization is due to reduced interfacial energy from solute segregation at the θ′/matrix interface. Al_3Sc stability is attributed to a Zr-rich shell around a Sc-rich core, which acts as a diffusion barrier that slows solute exchange and particle growth during thermal exposure. These findings highlight interfacial segregation as a promising strategy to stabilise dispersoid-strengthening in Al alloys. Li et al. [238] performed compressive creep tests at 300°C, 50 MPa for 100 h on Al-Cu 224 cast alloys micro-alloyed with Zr, V, and Sc to assess the stability of θ″/θ′-Al_2Cu precipitates and $L1_2$-Al_3TE dispersoids. Under T7A heat-treatment, the Sc and Zr containing alloy showed the lowest minimum creep rate (2.8×10^{-9} s^{-1}), compared to the base alloy (2.0×10^{-7} s^{-1}). Microstructural analysis attributed the good creep resistance to a high density of Al_3(Sc, Zr) dispersoids combined with finer θ′ precipitates that hindered dislocation motion, yielding a small creep strain of 0.007 after 100 h. In line with Gao et al. [34], this work also highlights interfacial segregation as a key mechanism for second-phase strengthening.

Bahl et al. [239] examined the influence of grain-boundary θ-Al_2Cu in Al-Cu-Mn-Zr alloys containing 6 wt.% and 9 wt.% Cu via tensile and compressive creep at 300°C. The θ phase at boundaries increased from ~0.7% in the 6Cu alloy to about 6% in the 9Cu alloy. Despite the higher fraction, the 9Cu alloy failed more rapidly in tensile creep above 20 MPa, while compressive creep rates between 110-115 MPa were similar for both compositions. Post-creep microstructures (Figure 2.50) showed that larger intergranular θ particles in the 9Cu alloy promoted cavity formation along grain boundaries under tension, leading to premature failure. Recent studies on Al-Cu alloys indicate that the addition of Ni demonstrates the need of minor alloying additions to enhance ductility and stabilize θ' precipitates at elevated temperatures. Notably, the combined addition of Zr and Sc, playing a significant role, offers a promising strategy for future high-temperature Al-Cu alloy development. The steady state creep rates of different reported Al-Cu based alloys are listed in Table 2.7, alongside the second phases associated with improved creep resistance.

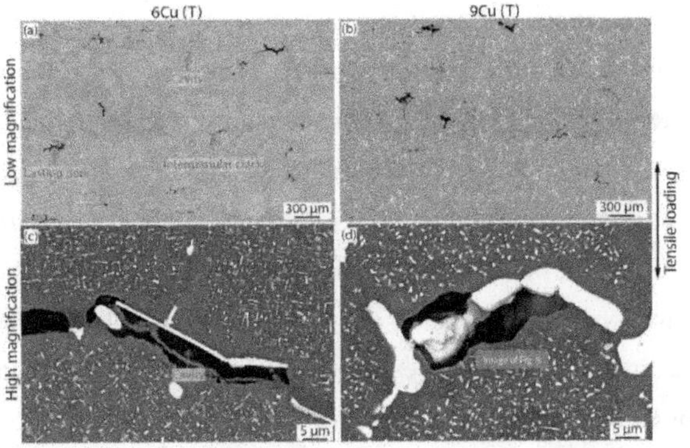

Figure 2.50 BSE-SEM images of cross-sections from failed tensile creep specimens of (a, c) Al-5.6Cu-0.39Mn-0.16Zr-0.06Fe-0.07Si and (b, d) Al-8.5Cu-0.36Mn-0.16Zr-0.07Fe-0.05Si alloys reveal creep cavities, intergranular cracking, and casting porosity [239].

Table 2.7 Creep parameters of Al-Cu alloys tested at 300 °C under 50 MPa.

Alloy composition (wt.%)	Minimum creep rate (s^{-1})		Key microstructure	Ref.
	Tensile	Compressive		
Al-4.9Cu-0.13Mg-0.38Mn-0.13Fe-0.12Si-0.11Ti	-	2.2×10^{-7}	θ', θ''	
Al-4.9Cu-0.11Mg-0.36Mn-0.13Fe-0.1Si-0.1Ti-0.18Zr	-	2.8×10^{-8}	θ', θ'', Al_3Zr	
Al-4.8Cu-0.13Mg-0.38Mn-0.13Fe-0.1Si-0.1Ti-0.25V	-	8.4×10^{-8}	θ', θ''	[238]
Al-4.7Cu-0.12Mg-0.35Mn-0.13Fe-0.08Si-0.09Ti-0.15Zr-0.13V	-	1.3×10^{-8}	θ', θ''	
Al-4.4Cu-0.11Mg-0.33Mn-0.12Fe-0.09Si-0.08Ti-0.15Zr-0.16Sc	-	5.5×10^{-9}	θ', θ'', Al_3(Sc, Zr)	
Al-5.6Cu-0.39Mn-0.16Zr-0.06Fe-0.07Si	7.0×10^{-8}	6.0×10^{-8}	θ', θ	[239]
Al-8.5Cu-0.36Mn-0.16Zr-0.07Fe-0.05Si	2.0×10^{-8}	6.0×10^{-9}	θ', θ	
Al-4.8Cu-0.19Mn-0.17Zr-0.09Fe-0.05Si-1.2Ni-0.26Co-0.21Ti	3.0×10^{-8}	-	θ', GB intermetallic (Al_7Cu_4Ni and Al_9Co_2)	
Al-6.4Cu-0.19Mn-0.13Zr-0.1Fe-0.05Si-0.09Ti	2.8×10^{-7}	-	θ', θ	[236]
Al-2.5Cu-0.3Sc	2.0×10^{-7}	-	θ', Al_3Sc	[75]

2.18 Summary and Research Gaps

A detailed literature review has been undertaken on the development of heat-resistant aluminium alloys, including its developmental history, criteria for developing heat-resistant alloys, aerospace grade alloys and their current status, microstructure tuning strategies and manufacturing routes. The aerospace-grade wrought materials; Al-Zn-Mg-Cu series AA7075 exhibits very high tensile strength at ambient temperature (~510 MPa) but experiences a sharp strength decline at high temperatures (>150°C) compared to other series of Al alloys. However, 2219 exhibits good heat resistance upto 200°C among Al alloys but suffers from strength deterioration above 200°C, limiting its applications under extreme conditions. However, inconsistencies have been identified in alloy compositions, aging conditions, operating temperature ranges, manufacturing processing methods and high-temperature characteristics. None of the reported alloys have achieved satisfactory thermal stability at 300°C, leaving industries still in need of lightweight, heat-resistant Al-based alloys. These challenges are summarized below:

1. **Limited in-situ studies on real-time failure mechanisms in multiphase Al-Zn-Mg-Cu alloys**: Limited attention has been paid to the in-situ study of the failure mechanisms in multiphase Al-Zn-Mg-Cu alloys at the micrometre scale. This gap is particularly relevant for 7075 alloys, where heterogeneities like grain boundaries, precipitates, and intermetallics cause complex failure modes (e.g., intergranular vs. transgranular fracture), and in-situ SEM is underutilized for observing these in real time.
2. **Optimization gaps in microstructure refinement for ambient and elevated-temp properties in both 7075Al and 2219Al alloys**: There is a gap in optimizing microalloying strategies, such as using grain nucleants like Al_3TM for ambient strength and elevated-temperature durability, to address processing inconsistencies, resulting in unbalanced room and high-temperature properties and necessitating research on hybrid microalloying approaches for enhanced thermal stability.
3. **Insufficient exploration on multiple solute segregation at precipitate/matrix interfaces of 2219Al alloy:** There is a notable research gap in exploring multiple solute segregation at θ' precipitate/matrix interfaces in 2219 Al alloy, where θ' acts as the primary strengthening phase but coarsens rapidly above 200°C, with extensive studies on single solutes to inhibit coarsening via $L1_2$ phases, yet limited investigations into synergistic multi-solute effects that could enhance interface coherency, reduce diffusivity, and improve thermal stability as per the LSW model.
4. **Inadequate creep resistance and deformation behavior at 300°C**: There is a significant research gap in the creep resistance and deformation behavior of heat-resistant Al alloys at 300°C and above, with limited data on tensile strength and tensile creep properties, and insufficient exploration of the combined impact of grain boundary

intermetallics, $L1_2$ dispersoids, and multiple solute segregation at precipitate/matrix interfaces within a single alloy system on high-temperature deformation mechanisms.

5. **Unexplored application of arc surface re-melting to micro-alloyed 2219Al alloy**: Arc surface remelting (ASR) is a cost-effective alternative to laser-based methods for enhancing microstructure and strength in Al alloys, its specific effects on Al-Cu alloys modified with micro-alloying elements like Ni, Sc, Ti, and Zr, particularly their high-temperature behaviour, have remained largely unexplored prior to this study.

This PhD thesis directly targets these research gaps. The findings are expected to advance the understanding of the strategies of development of heat-resistant aluminium alloys.

References

[1] D. Rabinovich, The allure of aluminium, Nat. Chem. 5 (2013) 76. https://doi.org/10.1038/nchem.1535

[2] Website: www.en.wikipedia.org/wiki/Alfred Gilbert.

[3] I.J. Polmear, Aluminium alloys-a century of age hardening, Mater. Forum 28 (2004) 13.

[4] S. Ranganathan, K. Chattopadhyay, Advanced aluminium alloys via rapid solidification, Aluminum India VIII (1991).

[5] S.J. Matthews, Thermal stability of solid solution strengthened high performance alloys, Superalloys: Metallurgy and Manufacture (1976) 215-226.

[6] F. Czerwinski, Thermal stability of aluminum alloys, Materials 13 (2020) 3441. https://doi.org/10.3390/ma13153441.

[7] E.J. Lavernia, T.S. Srivatsan, F.A. Mohamed, Strength, deformation, fracture behaviour and ductility of aluminium-lithium alloys, J. Mater. Sci. 25 (1990) 1137-1158. https://doi.org/10.1007/BF00585420.

[8] S.S. Li, X. Yue, Q.Y. Li, H.L. Peng, B.X. Dong, T.S. Liu, H.Y. Yang, J. Fan, S.L. Shu, F. Qiu, Q.C. Jiang, Development and applications of aluminum alloys for aerospace industry, J. Mater. Res. Technol. 27 (2023) 944-983. https://doi.org/10.1016/j.jmrt.2023.09.274.

[9] G. Dieter, Mechanical Metallurgy, McGraw-Hill Inc., New York, NY, USA, 1976.

[10] J. Friedel, Dislocation in Crystals, Pergamon Press, London, UK, 1967.

[11] J.C. Williams, E.A. Starke Jr, Progress in structural materials for aerospace systems, Acta Mater. 51 (2003) 5775-5799. https://doi.org/10.1016/j.actamat.2003.08.023.

[12] K.E. Knipling, D.C. Dunand, D.N. Seidman, Criteria for developing castable, creep-resistant aluminum-based alloys–A review, Int. J. Mater. Res. 97 (2022) 246-265. https://doi.org/10.1515/ijmr-2006-0042.

[13] K. Marthinsen, E. Nes, A general model for metal plasticity, Mater. Sci. Eng. A 234 (1997) 1095-1098. https://doi.org/10.1016/S0921-5093(97)00349-3.

[14] E.L. Rooy, Introduction to aluminum and aluminum alloys, Properties and Selection: Nonferrous Alloys and Special-Purpose Materials (1990) 3-14. https://doi.org/10.31399/asm.hb.v02.a0001057.

[15] D. Hull, D.J. Bacon, Introduction to dislocations, Elsevier, 2011.

[16] G.I. Taylor, The mechanism of plastic deformation of crystals. Part I.-Theoretical, Proc. R. Soc. London, Ser. A 145 (1934) 362-387. https://doi.org/10.1098/rspa.1934.0106.

[17] Argon, Strengthening mechanisms in crystal plasticity, OUP Oxford, 2007.

[18] J. Campbell, Complete casting handbook: metal casting processes, metallurgy, techniques and design, Butterworth-Heinemann, 2015.

[19] D.A. Porter, K.E. Easterling, M.Y. Sherif, Phase transformations in metals and alloys, CRC Press, Boca Raton, 2021.

[20] E. Balducci, L. Ceschini, S. Messieri, S. Wenner, R. Holmestad, Thermal stability of the lightweight 2099 Al-Cu-Li alloy: Tensile tests and microstructural investigations after overaging, Mater. Des. 119 (2017) 54-64. https://doi.org/10.1016/j.matdes.2017.01.058.

[21] H.A. Calderon, P.W. Voorhees, J.L. Murray, G. Kostorz, Ostwald ripening in concentrated alloys, Acta Metall. Mater. 42 (1994) 991-1000. https://doi.org/10.1016/0956-7151(94)90293-3.

[22] J.D. Boyd, R. Nicholson, The coarsening behaviour of θ″ and θ′ precipitates in two Al-Cu alloys, Acta Metall. 19 (1971) 1379-1391. https://doi.org/10.1016/0001-6160(71)90076-9.

[23] B. Noble, Theta-prime precipitation in aluminium-copper-cadmium alloys, Acta Metall. 16 (1968) 393-401. https://doi.org/10.1016/0001-6160(68)90026-6.

[24] V. Vaithyanathan, C. Wolverton, L.Q. Chen, Multiscale modeling of θ′ precipitation in Al-Cu binary alloys, Acta Mater. 52 (2004) 2973-2987. https://doi.org/10.1016/j.actamat.2004.03.001.

[25] H. Liu, I. Papadimitriou, F.X. Lin, J. LLorca, Precipitation during high temperature aging of Al-Cu alloys: A multiscale analysis based on first principles calculations, Acta Mater. 167 (2019) 121-135. https://doi.org/10.1016/j.actamat.2019.01.024.
[26] H. Liu, B. Bellón, J. Llorca, Multiscale modelling of the morphology and spatial distribution of θ′ precipitates in Al-Cu alloys, Acta Mater. 132 (2017) 611-626. https://doi.org/10.1016/j.actamat.2017.04.042.
[27] K. Kim, A. Roy, M.P. Gururajan, C. Wolverton, P.W. Voorhees, First-principles/Phase-field modeling of θ′ precipitation in Al-Cu alloys, Acta Mater. 140 (2017) 344-354. https://doi.org/10.1016/j.actamat.2017.08.046.
[28] M. Mantina, S.L. Shang, Y. Wang, L.Q. Chen, Z.K. Liu, 3 d transition metal impurities in aluminum: a first-principles study, Phys. Rev. B 80 (2009) 184111. https://doi.org/10.1103/PhysRevB.80.184111.
[29] G. Neumann, C. Tuijn, Self-Di_usion and Impurity Di_usion in Pure Metals: Handbook of Experimental Data, Pergamon Materials Series 14, Pergamon, Amsterdam, The Netherlands, 2009.
[30] G.M. Hood, The diffusion of iron in aluminium, Philos. Mag. 21 (1970) 305-328. https://doi.org/10.1080/14786437008238419.
[31] S.P. Murarka, R.P. Agarwala, Diffusion of rare earth elements in aluminum (No. BARC-368), Bhabha Atomic Research Centre, Bombay (India), 1968. https://www.osti.gov/biblio/4808164.
[32] J.E. Saal, C. Wolverton, Thermodynamic stability of Mg-based ternary long-period stacking ordered structures, Acta Mater. 68 (2014) 325-338. https://doi.org/10.1016/j.actamat.2013.10.055.
[33] Website: www.oqmd.org- open quantum materials database.
[34] R.A. Karnesky, D.N. Seidman, D. Dunand, Creep of Al-Sc microalloys with rare-earth element additions, Mater. Sci. Forum 519 (2006) 1035-1040. https://doi.org/10.4028/www.scientific.net/MSF.519-521.1035.
[35] J. Røyset, N. Ryum, Scandium in aluminium alloys, Int. Mater. Rev. 50 (2005) 19-44. https://doi.org/10.1179/174328005X14311.
[36] V.V. Zakharov, T.D. Rostova, Effect of scandium, transition metals, and admixtures on strengthening of aluminum alloys due to decomposition of the solid solution, Met. Sci. Heat Treat. 49 (2007) 435-442. https://doi.org/10.1007/s11041-007-0082-x.
[37] D.N. Seidman, E.A. Marquis, D.C. Dunand, Precipitation strengthening at ambient and elevated temperatures of heat-treatable Al (Sc) alloys, Acta Mater. 50 (2002) 4021-4035.

https://doi.org/10.1016/S1359-6454(02)00201-X.

[38] J. Royset, Scandium in aluminium alloys: physical metallurgy, properties and applications, Metall. Sci. Technol. 25 (2007). https://www.fracturae.com/index.php/MST/article/view/1126.

[39] E.A. Marquis, D.N. Seidman, Nanoscale structural evolution of Al_3Sc precipitates in Al (Sc) alloys, Acta Mater. 49 (2001) 1909-1919. https://doi.org/10.1016/S1359-6454(01)00116-1

[40] S. Lathabai, Fundamentals of Aluminium Metallurgy, Woodhead Publishing, U.S., 2018, pp. 47-92. https://doi.org/10.1016/B978-0-08-102063-0.00002-3.

[41] B. Ahuja, M. Karg, K.Y. Nagulin, M. Schmidt, Fabrication and characterization of high strength Al-Cu alloys processed using laser beam melting in metal powder bed, Phys. Procedia 56 (2014) 135-146. https://doi.org/10.1016/j.phpro.2014.08.156.

[42] D. Bayoumy, W. Kan, X. Wu, Y. Zhu, A. Huang, The latest development of Sc-strengthened aluminum alloys by laser powder bed fusion, J. Mater. Sci. Technol. 149 (2023) 1-17. https://doi.org/10.1016/j.jmst.2022.11.028.

[43] S. Costa, H. Puga, J. Barbosa, A.M.P. Pinto, The effect of Sc additions on the microstructure and age hardening behaviour of as cast Al–Sc alloys, Mater. Des. 42 (2012) 347-352. https://doi.org/10.1016/j.matdes.2012.06.019.

[44] A.F. Norman, P.B. Prangnell, R.S. McEwen, The solidification behaviour of dilute aluminium-scandium alloys, Acta Mater. 46 (1998) 5715-5732. https://doi.org/10.1016/S1359-6454(98)00257-2.

[45] P. Rometsch, Q. Jia, K.V. Yang, X. Wu, Additive Manufacturing for the Aerospace Industry, Elsevier, Netherlands, 2019, pp. 301-325. DOI:10.1016/C2017-0-00712-7.

[46] Y. Harada, D.C. Dunand, Thermal expansion of al3sc and al3 (sc0. 75x0. 25), Scr. Mater. 48 (2003) 219-222. https://doi.org/10.1016/S1359-6462(02)00428-1.

[47] A.B. Spierings, K. Dawson, M. Voegtlin, F. Palm, P.J. Uggowitzer, Microstructure and mechanical properties of as-processed scandium-modified aluminium using selective laser melting, CIRP Ann. 65 (2016) 213-216. https://doi.org/10.1016/j.cirp.2016.04.057.

[48] V.G. Davydov, T.D. Rostova, V.V. Zakharov, Y.A. Filatov, V.I. Yelagin, Scientific principles of making an alloying addition of scandium to aluminium alloys, Mater. Sci. Eng. A 280 (2000) 30-36. https://doi.org/10.1016/S0921-5093(99)00652-8.

[49] Y.A. Filatov, V.I. Yelagin, V.V. Zakharov, New Al–Mg–Sc alloys, Mater. Sci. Eng. A 280 (2000) 97-101. https://doi.org/10.1016/S0921-5093(99)00673-5.

[50] J. Murray, A. Peruzzi, J.P. Abriata, The Al-Zr (aluminum-zirconium) system, J. Phase Equilib. 13 (1992) 277-291. https://doi.org/10.1007/BF02667556.

[51] K.E. Knipling, D.C. Dunand, D.N. Seidman, Precipitation evolution in Al-Zr and Al-Zr-Ti alloys during isothermal aging at 375-425°C, Acta Mater. 56 (2008) 114-127. https://doi.org/10.1016/j.actamat.2007.09.004.

[52] K.E. Knipling, D.C. Dunand, D.N. Seidman, Precipitation evolution in Al-Zr and Al-Zr-Ti alloys during aging at 450-600°C, Acta Mater. 56 (2008) 1182-1195. https://doi.org/10.1016/j.actamat.2007.11.011.

[53] Y.D. Sui, Q.D. Wang, Application research and development of cast heat-resistant aluminum alloys in engines, Mater. Rev. 29 (2015) 14-19. DOI:10.1179/1743280413Y.0000000027.

[54] S. Roy, S. Roy, New-Age Al-Cu-Mn-Zr (ACMZ) Alloy for High Temperature-High Strength Applications: A Review, Aluminium Alloys-Design and Development of Innovative Alloys, Manufacturing Processes and Applications (2022) 13. DOI:10.5772/intechopen.104533.

[55] S. Wang, C. Fan, Crystal structures of Al_2Cu revisited: understanding existing phases and exploring other potential phases, Metals 9 (2019) 1037. https://doi.org/10.3390/met9101037.

[56] Z. Shen, Q. Ding, C. Liu, J. Wang, H. Tian, J. Li, Z. Zhang, Atomic-scale mechanism of the $\theta'' \rightarrow \theta'$ phase transformation in Al-Cu alloys, J. Mater. Sci. Technol. 33 (2017) 1159-1164. https://doi.org/10.1016/j.jmst.2016.08.031.

[57] D.A. Porter, K.E. Easterling, Phase transformations in metals and alloys (revised reprint), CRC Press, 2009. https://doi.org/10.1201/9781439883570.

[58] M.F. Chisholm, D. Shin, G. Duscher, M.P. Oxley, L.F. Allard, J.D. Poplawsky, A. Shyam, Atomic structures of interfacial solute gateways to θ' precipitates in Al-Cu alloys, Acta Mater. 212 (2021) 116891. https://doi.org/10.1016/j.actamat.2021.116891.

[59] Z.K. Zhang, Y. Che, S.Q. Men, X.M. Chen, Aging precipitation phase of Al-Cu-Mn high-strength aluminum alloy, Spec-cast Non-ferr Alloys 34 (2014) 1114-1116. https://doi.org/10.1007/s10853-024-09538-z.

[60] B. Lin, S. Xia, H. Li, Z. Lou, K. Liu, W. Zhang, Improved creep resistance of Al-Cu-Mn-Fe-Ni alloys through squeeze casting, Mater. Charact. 158 (2019) 109935. https://doi.org/10.1016/j.matchar.2019.109935.

[61] J. Chen, H. Liao, Y. Wu, H. Li, Contributions to high temperature strengthening from three types of heat-resistant phases formed during solidification, solution treatment and

ageing treatment of Al-Cu-Mn-Ni alloys respectively, Mater. Sci. Eng. A 772 (2020) 138819.https://doi.org/10.1016/j.msea.2019.138819.

[62] Y.Q. Chen, S.P. Pan, B.X. Zhu, X. Liu, W.H. Liu, S.W. Tang, The evolution of orientation relationships during the transformation of a twin-free T-particle to tenfold T-twins in an Al alloy during homogenisation, Mater. Charact. 141 (2018) 59-73. https://doi.org/10.1016/j.matchar.2018.03.029.

[63] Z. Shen, C. Liu, Q. Ding, S. Wang, X. Wei, L. Chen, J. Li, Z. Zhang, The structure determination of $Al_{20}Cu_2Mn_3$ by near atomic resolution chemical mapping, J. Alloys Compd. 601 (2014) 25-30. https://doi.org/10.1016/j.jallcom.2014.02.125.

[64] M. Cooper, K. Robinson, The crystal structure of the ternary alloy α(AlMnSi), Acta Crystallogr. 20 (1966) 614-617. https://doi.org/10.1107/S0365110X6600149X.

[65] L.F. Mondolfo, Aluminum alloys: structure and properties, Elsevier, 2013.

[66] G.H. Wu, Y.J. Guo, F.Z. Qi, S. Zhang, Y.X. Wang, X. Tong, L. Zhang, Role of alloying and heat treatment on microstructure and mechanical properties of cast Al-Li alloys: A review, China Foundry 21 (2024) 445-460. https://doi.org/10.1007/s41230-024-4062-1.

[67] C.L. Fu, Electronic, elastic, and fracture properties of trialuminide alloys: Al_3Sc and Al_3Ti, J. Mater. Res. 5 (1990) 971-979. DOI: https://doi.org/10.1557/JMR.1990.0971.

[68] M. Rakesh, N. Srivastava, S. Bhagavath, S. Karagadde, Role of In Situ Formed (Al_3Zr+ Al_3Ti) Particles on Nucleation of Primary Phase in Al-5 wt.% Cu Alloy, J. Mater. Eng. Perform. 33 (2024) 6488-6498. https://doi.org/10.1007/s11665-023-08417-z.

[69] A. Shyam, S. Roy, D. Shin, J.D. Poplawsky, L.F. Allard, Y. Yamamoto, J.R. Morris, B. Mazumder, J.C. Idrobo, A. Rodriguez, T.R. Watkins, Elevated temperature microstructural stability in cast AlCuMnZr alloys through solute segregation, Mater. Sci. Eng. A 765 (2019) 138279. https://doi.org/10.1016/j.msea.2019.138279.

[70] S. Bahl, L. Xiong, L.F. Allard, R.A. Michi, J.D. Poplawsky, A.C. Chuang, D. Singh, T.R. Watkins, D. Shin, J.A. Haynes, A. Shyam, Aging behavior and strengthening mechanisms of coarsening resistant metastable θ' precipitates in an Al-Cu alloy, Mater. Des. 198 (2021) 109378. https://doi.org/10.1016/j.matdes.2020.109378.

[71] J.D. Poplawsky, B.K. Milligan, L.F. Allard, D. Shin, P. Shower, M.F. Chisholm, A. Shyam, The synergistic role of Mn and Zr/Ti in producing θ'/$L1_2$ co-precipitates in Al-Cu alloys, Acta Mater. 194 (2020) 577-586. https://doi.org/10.1016/j.actamat.2020.05.043.

[72] X. Su, Y. Lei, Y. Chen, H. Qu, Z. Qi, G. Zheng, X. Liu, H. Xiang, G. Chen, Precipitating thermally reinforcement phase in aluminum alloys for enhanced strength at 400°C, J. Mater. Sci. Technol. 172 (2024) 71-82. https://doi.org/10.1016/j.jmst.2023.06.038.

[73] S. Srinivasan, P.B. Desch, R.B. Schwarz, Metastable phases in the Al sub 3 X (X= Ti, Zr, and Hf) intermetallic system, Scr. Metall. 25 (1991). https://www.osti.gov/biblio/5879598.

[74] P. Shower, J. Poplawsky, S. Bahl, A. Shyam, The role of Si in determining the stability of the θ′ precipitate in Al-Cu-Mn-Zr alloys, J. Alloys Compd. 862 (2021) 158152. https://doi.org/10.1016/j.jallcom.2020.158152.

[75] Y.H. Gao, C. Yang, J.Y. Zhang, L.F. Cao, G. Liu, J. Sun, E. Ma, Stabilizing nanoprecipitates in Al-Cu alloys for creep resistance at 300 C, Mater. Res. Lett. 7 (2019) 18-25. https://doi.org/10.1080/21663831.2018.1546773

[76] Y.H. Gao, L.F. Cao, C. Yang, J.Y. Zhang, G. Liu, J. Sun, Co-stabilization of θ′-Al_2Cu and Al_3Sc precipitates in Sc-microalloyed Al-Cu alloy with enhanced creep resistance, Mater. Today Nano 6 (2019) 100035. https://doi.org/10.1016/j.mtnano.2019.100035.

[77] L. Jiang, B. Rouxel, T. Langan, T. Dorin, Coupled segregation mechanisms of Sc, Zr and Mn at θ′ interfaces enhances the strength and thermal stability of Al-Cu alloys, Acta Mater. 206 (2021) 116634. https://doi.org/10.1016/j.actamat.2021.116634.

[78] J. Rakhmonov, K. Liu, L. Pan, F. Breton, X.G. Chen, Enhanced mechanical properties of high-temperature-resistant Al-Cu cast alloy by microalloying with Mg, J. Alloys Compd. 827 (2020) 154305. https://doi.org/10.1016/j.jallcom.2020.154305.

[79] S. Mondol, S. Kashyap, S. Kumar, K. Chattopadhyay, Improvement of high temperature strength of 2219 alloy by Sc and Zr addition through a novel three-stage heat treatment route, Mater. Sci. Eng. A 732 (2018) 157-166. https://doi.org/10.1016/j.msea.2018.07.003.

[80] C. Yang, P. Zhang, D. Shao, R.H. Wang, L.F. Cao, J.Y. Zhang, G. Liu, B.A. Chen, J. Sun, The influence of Sc solute partitioning on the microalloying effect and mechanical properties of Al-Cu alloys with minor Sc addition, Acta Mater. 119 (2016) 68-79. https://doi.org/10.1016/j.actamat.2016.08.013.

[81] U. Bansal, M.P. Singh, S. Mondol, S.K. Sinha, S.K. Makineni, A. Paul, K. Chattopadhyay, The interplay of precipitation of ordered compounds and interfacial segregation in Al-Cu-Hf-Si alloys for high-temperature strength, Acta Mater. 240 (2022) 118355. https://doi.org/10.1016/j.actamat.2022.118355.

[82] U. Bansal, M.P. Singh, S.K. Sinha, D.K. Sahu, S. Mondol, S.K. Makineni, A. Paul, K. Chattopadhyay, Strength and stability through variable micro segregation behaviour of Ta and Zr solutes at intermetallic interfaces in Al-Cu alloys, Acta Mater. 259 (2023) 119254.https://doi.org/10.1016/j.actamat.2023.119254.

[83] X.Y. Liu, Q.L. Pan, Z.L. Lu, S. Cao, Y. He, W. Li, Thermal stability of Al-Cu-Mg-Ag heat resistant alloy, Chin. J. Nonferrous Met. 21 (2011) 1244-1251.

[84] Y.H. Gao, P.F. Guan, R. Su, H.W. Chen, C. Yang, C. He, L.F. Cao, H. Song, J.Y. Zhang, X.F. Zhang, G. Liu, Segregation-sandwiched stable interface suffocates nanoprecipitate coarsening to elevate creep resistance, Mater. Res. Lett. 8 (2020) 446-453. https://doi.org/10.1080/21663831.2020.1799447.

[85] Y.H. Gao, L.F. Cao, J. Kuang, H. Song, G. Liu, J.Y. Zhang, J. Sun, Solute repositioning to tune the multiple microalloying effects in an Al-Cu alloy with minor Sc, Fe and Si addition, Mater. Sci. Eng. A 803 (2021) 140509. https://doi.org/10.1016/j.msea.2020.140509.

[86] J.G. Kaufman (Ed.), Properties of aluminum alloys: tensile, creep, and fatigue data at high and low temperatures, ASM International, 1999.

[87] R. Li, N. Takata, A. Suzuki, M. Kobashi, Design of heat-resistant Al-Mg-Zn-Cu-Ni quinary alloy: controlling intermetallic phases and mechanical performance at elevated temperature, Mater. Sci. Eng. A 857 (2022) 144055. https://doi.org/10.1016/j.msea.2022.144055.

[88] K. Yu, W. Li, S. Li, J. Zhao, Mechanical properties and microstructure of aluminum alloy 2618 with Al_3(Sc, Zr) phases, Mater. Sci. Eng. A 368 (2004) 88-93. https://doi.org/10.1016/j.msea.2003.09.092.

[89] M.T. Sun, Research on the composition and properties of A201 aluminum alloy, Ph.D. Dissertation, Shenyang Foundry Research Institute, 2011. https://doi.org/10.1007/s10853-024-09538-z.

[90] Z. Yang, M. Zhang, L. Fan, J. Bai, X. Chen, J. Zhang, Z. Li, R. Guan, Towards high mechanical performance Al-Cu-Mg-Fe-Ni alloy: influence of composition, solution treatment and aged process on microstructural evolution and mechanical properties, J. Mater. Res. Technol. 23 (2023) 2054-2064. https://doi.org/10.1016/j.jmrt.2023.01.166.

[91] B. Lin, W. Zhang, X. Zheng, Y. Zhao, Z. Lou, W. Zhang, Developing high performance mechanical properties at elevated temperature in squeeze cast Al-Cu-Mn-Fe-Ni alloys, Mater. Charact. 150 (2019) 128-137. https://doi.org/10.1016/j.matchar.2019.01.022.

[92] I.J. Polmear, A trace element effect in alloys based on the aluminium-zinc-magnesium system, Nature 186 (1960) 303-304. https://doi.org/10.1038/186303a0.

[93] Y. Dai, L. Yan, J. Hao, Review on micro-alloying and preparation method of 7xxx series aluminum alloys: progresses and prospects, Materials 15 (2022) 1216. https://doi.org/10.3390/ma15031216.

[94] L.L. Rokhlin, T.V. Dobatkina, N.R. Bochvar, E.V. Lysova, Investigation of phase equilibria in alloys of the Al-Zn-Mg-Cu- Zr-Sc system, J. Alloys Compd. 367 (2004) 10-16. https://doi.org/10.1016/j.jallcom.2003.08.003.

[95] X.M. Zhang, Y.L. Deng, China's Strategic Emerging Industry-New Alloy Materials, China Railway Publishing House, Beijing, 2018, pp. 25-26.

[96] S.J. Won, H. So, L. Kang, S.J. Oh, K.H. Kim, Development of a high-strength Al-Zn-Mg-Cu-based alloy via multi-strengthening mechanisms, Scr. Mater. 205 (2021) 114216. https://doi.org/10.1016/j.scriptamat.2021.114216.

[97] P. Priya, D.R. Johnson, M.J. Krane, Precipitation during cooling of 7XXX aluminum alloys, Comput. Mater. Sci. 139 (2017) 273-284. https://doi.org/10.1016/j.commatsci.2017.08.008.

[98] Y. Zhang, H.P. Li, W. Kang, X.M. Zhang, Aging microstructure evolution in high strength aluminum alloys and performance controlling, Chin. J. Nonferr. Met. 27 (2017) 1323-1336. DOI:10.19476/j.ysxb.1004.0609.2017.07.02.

[99] X.W. Yu, J.H. Chen, W.Q. Ming, X.B. Yang, T.T. Zhao, R.H. Shen, Y.T. He, C.L. Wu, Revisiting the Hierarchical Microstructures of an Al-Zn-Mg Alloy Fabricated by Pre-deformation and Aging: X.-W. Yu et al, Acta Metall. Sin. (Engl. Lett.) 33 (2020) 1518-1526.https://doi.org/10.1007/s40195-020-01082-4.

[100] R. Bakhshi, M.H. Farshidi, S.A. Sajjadi, Strengthening of aluminium alloy 7005 through imposition of severe plastic deformation supplemented by different ageing treatments, Trans. Nonferrous Met. Soc. China 31 (2021) 2909-2921. https://doi.org/10.1016/S1003-6326(21)65702-3.

[101] T. Dursun, C. Soutis, Recent developments in advanced aircraft aluminium alloys, Mater. Des. (1980-2015) 56 (2014) 862-871. https://doi.org/10.1016/j.matdes.2013.12.002.

[102] A.P. Mouritz, Introduction to aerospace materials, Elsevier, 2012.

[103] H. Xiao, Z. Wang, J. Geng, C. Zhang, Y. Li, Q. Yang, M. Wang, D. Chen, Z. Li, H. Wang, Precipitation and crystallographic relationships of nanosized η/η' precipitates at S-Al interface in Al-Zn-Mg-Cu alloy, Scr. Mater. 214 (2022) 114643.

https://doi.org/10.1016/j.scriptamat.2022.114643.

[104] S.V. Senkova, O.N. Senkov, D.B. Miracle, Cryogenic and elevated temperature strengths of an Al-Zn-Mg-Cu alloy modified with Sc and Zr, Metall. Mater. Trans. A 37 (2006) 3569-3575. https://doi.org/10.1007/s11661-006-1051-5.

[105] P. Dai, X. Luo, Y. Yang, Z. Kou, B. Huang, J. Zang, J. Ru, High temperature tensile properties, fracture behaviors and nanoscale precipitate variation of an Al-Zn-Mg-Cu alloy, Prog. Nat. Sci.: Mater. Int. 30 (2020) 63-73. https://doi.org/10.1016/j.pnsc.2020.01.007.

[106] J. Zhao, S. Bai, T. Huang, J. Wang, H. Xie, D. Zeng, L. Luo, Effect of various aging treatment on thermal stability of a novel Al-Zn-Mg-Cu alloy for oil drilling, Mater. Sci. Eng. A 803 (2021) 140490. https://doi.org/10.1016/j.msea.2020.140490.

[107] D. Liu, D. Wu, Y. Wang, Z. Chen, C. Ge, Q. Zhao, F. Niu, G. Ma, Enhanced high-temperature mechanical properties of laser-arc hybrid additive manufacturing of Al-Zn-Mg-Cu alloy via microstructure control, J. Mater. Sci. Technol. 169 (2024) 220-234. https://doi.org/10.1016/j.jmst.2023.05.071.

[108] Y. Zou, L. Cao, X. Wu, C. Mou, S. Tang, Revealing the coarsening behavior of precipitates and its effect on the thermal stability in T′ and η′ dual-phase strengthened Al-Zn-Mg-Cu alloys, J. Mater. Sci. Technol. 220 (2025) 54-66. https://doi.org/10.1016/j.jmst.2024.09.010.

[109] N. Takata, M. Ishihara, A. Suzuki, M. Kobashi, Microstructure and strength of a novel heat-resistant aluminum alloy strengthened by T-$Al_6Mg_{11}Zn_{11}$ phase at elevated temperatures, Mater. Sci. Eng. A 739 (2019) 62-70. https://doi.org/10.1016/j.msea.2018.10.034.

[110] C.Y. Liu, G.B. Teng, Z.Y. Ma, L.L. Wei, B. Zhang, Y. Chen, Effects of Sc and Zr microalloying on the microstructure and mechanical properties of high Cu content 7xxx Al alloy, Int. J. Miner. Metall. Mater. 26 (2019) 1559-1569. https://doi.org/10.1007/s12613-019-1840-7.

[111] M. Jia, Z. Zheng, Z. Gong, Microstructure evolution of the 1469 Al-Cu-Li-Sc alloy during homogenization, J. Alloys Compd. 614 (2014) 131-139. https://doi.org/10.1016/j.jallcom.2014.06.033.

[112] X. Zeng, C. Ferguson, K. Sadayappan, S. Shankar, Effect of titanium levels on the hot tearing sensitivity and abnormal grain growth after T4 heat treatment of Al-Zn-Mg-Cu alloys, Int. J. Metalcast. 12 (2018) 457-468. https://doi.org/10.1007/s40962-018-0227-2.

[113] S.H. Lee, J.G. Jung, S.I. Baik, S.H. Park, M.S. Kim, Y.K. Lee, K. Euh, Effects of Ti addition on the microstructure and mechanical properties of Al-Zn-Mg-Cu-Zr alloy, Mater. Sci. Eng. A 801 (2021) 140437. https://doi.org/10.1016/j.msea.2020.140437.

[114] S.H. Kayani, J.G. Jung, M.S. Kim, K. Euh, Effect of cooling rate on precipitation behavior of Al-7.65 Zn-2.59 Mg-1.95 Cu alloy with minor elements of Zr and Ti, Met. Mater. Int. 26 (2020) 1079-1086. https://doi.org/10.1007/s12540-019-00385-1.

[115] S.H. Lee, S.H. Kayani, J.G. Jung, S.I. Baik, M.S. Kim, Y.K. Lee, K. Euh, Crystallographic characterization of $Al_{18}Mg_3Ti_2$ intermetallic phase in Al-Zn-Mg-Cu-Zr-Ti alloy, J. Alloys Compd. 844 (2020) 156173. https://doi.org/10.1016/j.jallcom.2020.156173.

[116] Q. Zhu, L. Cao, X. Wu, Y. Zou, M.J. Couper, Effect of Ag on age-hardening response of Al-Zn-Mg-Cu alloys, Mater. Sci. Eng. A 754 (2019) 265-268. https://doi.org/10.1016/j.msea.2019.03.090.

[117] P. Dumitraschkewitz, P.J. Uggowitzer, S.S. Gerstl, J.F. Löffler, S. Pogatscher, Size-dependent diffusion controls natural aging in aluminium alloys, Nat. Commun. 10 (2019) 4746. https://doi.org/10.1038/s41467-019-12762-w.

[118] V.V. Zakharov, Some problems of the use of aluminum-lithium alloys, Met. Sci. Heat Treat. 45 (2003) 49-54. https://doi.org/10.1023/A:1023904314848.

[119] G.L. Shneider, Stability of supersaturated solid solution of aluminum-lithium alloy 1470, Met. Sci. Heat Treat. 40 (1998) 294-298. https://doi.org/10.1007/BF02474896.

[120] V.V. Zakharov, Thermal stability of Al-Li alloys, Met. Sci. Heat Treat. 41 (1999) 39-43. https://doi.org/10.1007/BF02466271.

[121] A.J. McAlister, The Al-Li (aluminum-lithium) system, Bull. Alloy Phase Diagrams 3 (1982) 177-183. https://doi.org/10.1007/BF02892377.

[122] W. Yichang, T. Xing, Y. Guoqiang, W. Xiaodong, C. Lingfei, Research progress and prospects of the microstructures, properties, and forming techniques of Al-Li Alloy, Rare Met. Mater. Eng. 50 (2021) 1069-1083. DOI:10.1016/j.jmapro.2022.07.067.

[123] R.J. Rioja, J. Liu, The evolution of Al-Li base products for aerospace and space applications, Metall. Mater. Trans. A 43 (2012) 3325-3337. https://doi.org/10.1007/s11661-012-1155-z.

[124] T. Pasang, N. Symonds, S. Moutsos, R.J.H. Wanhill, S.P. Lynch, Low-energy intergranular fracture in Al-Li alloys, Eng. Fail. Anal. 22 (2012) 166-178. https://doi.org/10.1016/j.engfailanal.2012.01.006.

[125] J. Zhang, G. Wu, L. Zhang, X. Zhang, C. Shi, X. Tong, Addressing the strength-ductility trade-off in a cast Al-Li-Cu alloy-Synergistic effect of Sc-alloying and optimized artificial ageing scheme, J. Mater. Sci. Technol. 96 (2022) 212-225. https://doi.org/10.1016/j.jmst.2021.01.036.
[126] S. Kalyanam, A.J. Beaudoin, R.H. Dodds Jr, F. Barlat, Delamination cracking in advanced aluminum-lithium alloys-Experimental and computational studies, Eng. Fract. Mech. 76 (2009) 2174-2191. https://doi.org/10.1016/j.engfracmech.2009.06.010.
[127] B. Noble, S.J. Harris, S. Katsikis, K. Dinsdale, Low temperature thermal stability of quaternary Al-Li-Cu-Mg alloys, Mater. Sci. Forum 519 (2006) 209-214. https://doi.org/10.4028/www.scientific.net/MSF.519-521.209.
[128] R.J.H. Wanhill, Aerospace applications of aluminum-lithium alloys, Aluminum-lithium Alloys (2014) 503-535. https://doi.org/10.1016/B978-0-12-401698-9.00015-X.
[129] X. Yu, Z. Zhao, D. Shi, H. Dai, J. Sun, X. Dong, Enhanced high-temperature mechanical properties of Al-Cu-Li alloy through T1 coarsening inhibition and Ce-containing intermetallic refinement, Materials 12 (2019) 1521. https://doi.org/10.3390/ma12091521.
[130] D. Ortiz, J. Brown, M. Abdelshehid, P. DeLeon, R. Dalton, L. Mendez, J. Soltero, M. Pereira, M. Hahn, E. Lee, J. Ogren, The effects of prolonged thermal exposure on the mechanical properties and fracture toughness of C458 aluminum-lithium alloy, Eng. Fail. Anal. 13 (2006) 170-180. https://doi.org/10.1016/j.engfailanal.2004.10.008.
[131] M. Eddahbi, C.B. Thomson, F. Carreño, O.A. Ruano, Grain structure and microtexture after high temperature deformation of an Al-Li (8090) alloy, Mater. Sci. Eng. A 284 (2000) 292-300. https://doi.org/10.1016/S0921-5093(00)00728-0.
[132] X. Zheng, P. Luo, G. Yue, Y. Hu, Analysis of microstructure and high-temperature tensile properties of 2060 Al-Li alloy strengthened by laser shock peening, J. Alloys Compd. 860 (2021) 158539. https://doi.org/10.1016/j.jallcom.2020.158539.
[133] X. Zhang, H. Wang, C. Zou, Z. Wei, The evolution of microstructure and mechanical properties at elevated temperature of cast Al-Li-Cu-Mg alloys with Ni addition, J. Mater. Res. Technol. 9 (2020) 11069-11079. https://doi.org/10.1016/j.jmrt.2020.07.098.
[134] S. Zhang, F. Qi, L. Zhang, G. Wu, X. Tong, J. Zhan, Y. Wang, Y. Geng, Effect of aging treatment on high temperature mechanical properties of a cast Al-Li-Cu alloy, Mater. Sci. Eng. A 896 (2024) 146300. https://doi.org/10.1016/j.msea.2024.146300.
[135] S.C. Wang, M.J. Starink, Two types of S phase precipitates in Al-Cu-Mg alloys, Acta Mater. 55 (2007) 933-941. https://doi.org/10.1016/j.actamat.2006.09.015.

[136] Y. Wang, B. Sharma, Y. Xu, K. Shimizu, H. Fujihara, K. Hirayama, A. Takeuchi, M. Uesugi, G. Cheng, H. Toda, Switching nanoprecipitates to resist hydrogen embrittlement in high-strength aluminum alloys, Nat. Commun. 13 (2022) 6860. https://doi.org/10.1038/s41467-022-34628-4.

[137] M.R. Gazizov, A.N. Belyakov, R. Holmestad, M.Y. Gazizova, V.S. Krasnikov, P.A. Bezborodova, R.O. Kaibyshev, The coarsening behavior of strengthening particles in an Al-Cu-Mg-Ag alloy during creep, Mater. Sci. Eng. A 884 (2023) 145515. https://doi.org/10.1016/j.msea.2023.145515.

[138] S.A. Zhou, Z. Zhang, M. Li, D. Pan, H. Su, X. Du, P. Li, Y. Wu, Effect of Sc on microstructure and mechanical properties of as-cast Al-Mg alloys, Mater. Des. 90 (2016) 1077-1084. https://doi.org/10.1016/j.matdes.2015.10.132.

[139] S.H. Wu, H. Xue, C. Yang, P.M. Cheng, P. Zhang, J. Kuang, J.Y. Zhang, G. Liu, J. Sun, Effect of Si addition on the precipitation and mechanical/electrical properties of dilute Al-Zr-Sc alloys, Mater. Sci. Eng. A 812 (2021) 141150. https://doi.org/10.1016/j.msea.2021.141150.

[140] J. Liu, P. Yao, N. Zhao, C. Shi, H. Li, X. Li, D. Xi, S. Yang, Effect of minor Sc and Zr on recrystallization behavior and mechanical properties of novel Al-Zn-Mg-Cu alloys, J. Alloys Compd. 657 (2016) 717-725. https://doi.org/10.1016/j.jallcom.2015.10.122.

[141] M.J. Jones, F.J. Humphreys, Interaction of recrystallization and precipitation: The effect of Al_3Sc on the recrystallization behaviour of deformed aluminium, Acta Mater. 51 (2003) 2149-2159. https://doi.org/10.1016/S1359-6454(03)00002-8.

[142] Z. Yin, Q. Pan, Y. Zhang, F. Jiang, Effect of minor Sc and Zr on the microstructure and mechanical properties of Al-Mg based alloys, Mater. Sci. Eng. A 280 (2000) 151-155. https://doi.org/10.1016/S0921-5093(99)00682-6.

[143] E.A. Marquis, D.N. Seidman, D.C. Dunand, Effect of Mg addition on the creep and yield behavior of an Al-Sc alloy, Acta Mater. 51 (2003) 4751-4760. https://doi.org/10.1016/S1359-6454(03)00288-X.

[144] R.A. Karnesky, M.E. Van Dalen, D.C. Dunand, D.N. Seidman, Effects of substituting rare-earth elements for scandium in a precipitation-strengthened Al-0.08 at.% Sc alloy, Scr. Mater. 55 (2006) 437-440. https://doi.org/10.1016/j.scriptamat.2006.05.021.

[145] O. Beeri, D.C. Dunand, D.N. Seidman, Roles of impurities on precipitation kinetics of dilute Al-Sc alloys, Mater. Sci. Eng. A 527 (2010) 3501-3509. https://doi.org/10.1016/j.msea.2010.02.027.

[146] K.E. Knipling, R.A. Karnesky, C.P. Lee, D.C. Dunand, D.N. Seidman, Precipitation evolution in Al-0.1 Sc, Al-0.1 Zr and Al-0.1 Sc-0.1 Zr (at.%) alloys during isochronal aging, Acta Mater. 58 (2010) 5184-5195. https://doi.org/10.1016/j.actamat.2010.05.054.

[147] K.E. Knipling, D.N. Seidman, D.C. Dunand, Ambient-and high-temperature mechanical properties of isochronally aged Al-0.06 Sc, Al-0.06 Zr and Al-0.06 Sc-0.06 Zr (at.%) alloys, Acta Mater. 59 (2011) 943-954. https://doi.org/10.1016/j.actamat.2010.10.017.

[148] C. Booth-Morrison, Z. Mao, M. Diaz, D.C. Dunand, C. Wolverton, D.N. Seidman, Role of silicon in accelerating the nucleation of Al_3(Sc, Zr) precipitates in dilute Al-Sc-Zr alloys, Acta Mater. 60 (2012) 4740-4752. https://doi.org/10.1016/j.actamat.2012.05.036.

[149] N.Q. Vo, D.N. Seidman, D.C. Dunand, Effect of Si micro-addition on creep resistance of a dilute Al-Sc-Zr-Er alloy, Mater. Sci. Eng. A 734 (2018) 27-33. https://doi.org/10.1016/j.msea.2018.07.053.

[150] N.Q. Vo, D.C. Dunand, D.N. Seidman, Role of silicon in the precipitation kinetics of dilute Al-Sc-Er-Zr alloys, Mater. Sci. Eng. A 677 (2016) 485-495. https://doi.org/10.1016/j.msea.2016.09.065.

[151] N.Q. Vo, D.C. Dunand, D.N. Seidman, Improving aging and creep resistance in a dilute Al-Sc alloy by microalloying with Si, Zr and Er, Acta Mater. 63 (2014) 73-85. https://doi.org/10.1016/j.actamat.2013.10.008.

[152] C. Booth-Morrison, D.N. Seidman, D.C. Dunand, Effect of Er additions on ambient and high-temperature strength of precipitation-strengthened Al-Zr-Sc-Si alloys, Acta Mater. 60 (2012) 3643-3654. https://doi.org/10.1016/j.actamat.2012.02.030.

[153] C. Booth-Morrison, D.C. Dunand, D.N. Seidman, Coarsening resistance at 400°C of precipitation-strengthened Al-Zr-Sc-Er alloys, Acta Mater. 59 (2011) 7029-7042. https://doi.org/10.1016/j.actamat.2011.07.057.

[154] A. De Luca, D.C. Dunand, D.N. Seidman, Mechanical properties and optimization of the aging of a dilute Al-Sc-Er-Zr-Si alloy with a high Zr/Sc ratio, Acta Mater. 119 (2016) 35-42. https://doi.org/10.1016/j.actamat.2016.08.018.

[155] D. Erdeniz, W. Nasim, J. Malik, A.R. Yost, S. Park, A. De Luca, N.Q. Vo, I. Karaman, B. Mansoor, D.N. Seidman, D.C. Dunand, Effect of vanadium micro-alloying on the microstructural evolution and creep behavior of Al-Er-Sc-Zr-Si alloys, Acta Mater. 124 (2017) 501-512. https://doi.org/10.1016/j.actamat.2016.11.033.

[156] A. De Luca, D.C. Dunand, D.N. Seidman, Microstructure and mechanical properties of a precipitation-strengthened Al-Zr-Sc-Er-Si alloy with a very small Sc content, Acta Mater. 144 (2018) 80-91. https://doi.org/10.1016/j.actamat.2017.10.040.

[157] J.G. Jung, A.R. Farkoosh, D.N. Seidman, Microstructural and mechanical properties of precipitation-strengthened Al-Mg-Zr-Sc-Er-Y-Si alloys, Acta Mater. 257 (2023) 119167. https://doi.org/10.1016/j.actamat.2023.119167.

[158] E. Clouet, L. Laé, T. Épicier, W. Lefebvre, M. Nastar, A. Deschamps, Complex precipitation pathways in multicomponent alloys, Nat. Mater. 5 (2006) 482-488. https://doi.org/10.1038/nmat1652.

[159] A. Orthacker, G. Haberfehlner, J. Taendl, M.C. Poletti, B. Sonderegger, G. Kothleitner, Diffusion-defining atomic-scale spinodal decomposition within nanoprecipitates, Nat. Mater. 17 (2018) 1101-1107. https://doi.org/10.1038/s41563-018-0209-z.

[160] G. Ghosh, S. Vaynman, M. Asta, M.E. Fine, Stability and elastic properties of $L1_2$-(Al, Cu)$_3$(Ti, Zr) phases: Ab initio calculations and experiments, Intermetallics 15 (2007) 44-54. https://doi.org/10.1016/j.intermet.2006.03.003.

[161] R.K. Pan, H.C. Wang, T.T. Shi, X. Tian, B.Y. Tang, Thermal properties and thermoelasticity of $L1_2$ ordered Al_3RE (RE= Er, Tm, Yb, Lu) phases: a first-principles study, Mater. Des. 102 (2016) 100-105. https://doi.org/10.1016/j.matdes.2016.03.165.

[162] Y.H. Gao, L.F. Cao, J. Kuang, H. Song, J.Y. Zhang, G. Liu, J. Sun, Si-mediated reassembly of interfacially segregated Sc atoms in an Al-Cu-Sc alloy exposed to high-temperature creep, J. Alloys Compd. 845 (2020) 156266. https://doi.org/10.1016/j.jallcom.2020.156266.

[163] H. Xue, C. Yang, P. Zhang, S.H. Wu, G. Liu, J. Sun, Heat-resistant Al alloys: microstructural design and microalloying effect, J. Mater. Sci. 59 (2024) 9749-9767. https://doi.org/10.1007/s10853-023-09295-5.

[164] B.A. Chen, G. Liu, R.H. Wang, J.Y. Zhang, L. Jiang, J.J. Song, J. Sun, Effect of interfacial solute segregation on ductile fracture of Al-Cu-Sc alloys, Acta Mater. 61 (2013) 1676-1690. https://doi.org/10.1016/j.actamat.2012.11.043.

[165] J. Peng, S. Bahl, A. Shyam, J.A. Haynes, D. Shin, Solute-vacancy clustering in aluminum, Acta Mater. 196 (2020) 747-758. https://doi.org/10.1016/j.actamat.2020.06.062.

[166] Y.H. Gao, L.F. Cao, J. Kuang, J.Y. Zhang, G. Liu, J. Sun, Dual effect of Cu on the Al_3Sc nanoprecipitate coarsening, J. Mater. Sci. Technol. 37 (2020) 38-45. https://doi.org/10.1016/j.jmst.2019.07.035.

[167] C. Yang, P. Cheng, B. Chen, J. Zhang, G. Liu, J. Sun, Solute clusters-promoted strength-ductility synergy in Al-Sc alloy, J. Mater. Sci. Technol. 96 (2022) 325-331. https://doi.org/10.1016/j.jmst.2021.02.075.

[168] S.A. Dregia, P. Wynblatt, Equilibrium segregation and interfacial energy in multicomponent systems, Acta Metall. Mater. 39 (1991) 771-778. https://doi.org/10.1016/0956-7151(91)90277-8.

[169] A. Biswas, D.J. Siegel, C. Wolverton, D.N. Seidman, Precipitates in Al-Cu alloys revisited: Atom-probe tomographic experiments and first-principles calculations of compositional evolution and interfacial segregation, Acta Mater. 59 (2011) 6187-6204. https://doi.org/10.1016/j.actamat.2011.06.036.

[170] J.M. Rosalie, L. Bourgeois, Silver segregation to θ′(Al_2Cu)-Al interfaces in Al-Cu-Ag alloys, Acta Mater. 60 (2012) 6033-6041. https://doi.org/10.1016/j.actamat.2012.07.039.

[171] D. Shin, A. Shyam, S. Lee, Y. Yamamoto, J.A. Haynes, Solute segregation at the Al/θ′-Al_2Cu interface in Al-Cu alloys, Acta Mater. 141 (2017) 327-340. https://doi.org/10.1016/j.actamat.2017.09.020.

[172] Q. Lu, J. Wang, H. Li, S. Jin, G. Sha, J. Lu, L. Wang, B. Jin, X. Lan, L. Li, K. Li, Synergy of multiple precipitate/matrix interface structures for a heat resistant high-strength Al alloy, Nat. Commun. 14 (2023) 2959. https://doi.org/10.1038/s41467-023-38730-z.

[173] M.V. Petrik, Y.N. Gornostyrev, P.A. Korzhavyi, Segregation of alloying elements to stabilize θ′ phase interfaces in Al-Cu based alloys, Scr. Mater. 202 (2021) 114006. https://doi.org/10.1016/j.scriptamat.2021.114006.

[174] Y. Li, Y. Yang, Y. Wu, L. Wang, X. Liu, Quantitative comparison of three Ni-containing phases to the elevated-temperature properties of Al-Si piston alloys, Mater. Sci. Eng. A 527 (2010) 7132-7137. https://doi.org/10.1016/j.msea.2010.07.073.

[175] J.U. Rakhmonov, S. Bahl, A. Shyam, D.C. Dunand, Cavitation-resistant intergranular precipitates enhance creep performance of θ′-strengthened Al-Cu based alloys, Acta Mater. 228 (2022) 117788. https://doi.org/10.1016/j.actamat.2022.117788.

[176] H. Du, S. Zhang, B. Zhang, X. Tao, Z. Yao, N. Belov, S. van der Zwaag, Z. Liu, Ca-modified Al-Mg-Sc alloy with high strength at elevated temperatures due to a hierarchical microstructure, J. Mater. Sci. 56 (2021) 16145-16157. https://doi.org/10.1007/s10853-021-06310-5.

[177] S. Zhang, H. Du, Z. Yao, Z. Liu, Y. Zhu, L. Shuai, T. Huang, X. Huang, X. Tao, D.P. Mondal, T. Akopyan, Superior high temperature creep resistance of a cast Al-Mg-Ca-Sc

alloy with multi-scale hierarchical microstructures, Mater. Sci. Eng. A 850 (2022) 143533. https://doi.org/10.1016/j.msea.2022.143533.

[178] Y. Liu, R.A. Michi, D.C. Dunand, Cast near-eutectic Al-12.5 wt.% Ce alloy with high coarsening and creep resistance, Mater. Sci. Eng. A 767 (2019) 138440. https://doi.org/10.1016/j.msea.2019.138440.

[179] P. Pandey, S.K. Makineni, B. Gault, K. Chattopadhyay, On the origin of a remarkable increase in the strength and stability of an Al rich Al-Ni eutectic alloy by Zr addition, Acta Mater. 170 (2019) 205-217. https://doi.org/10.1016/j.actamat.2019.03.025.

[180] Z.C. Sims, O.R. Rios, D. Weiss, P.E. Turchi, A. Perron, J.R. Lee, T.T. Li, J.A. Hammons, M. Bagge-Hansen, T.M. Willey, K. An, High performance aluminum-cerium alloys for high-temperature applications, Mater. Horiz. 4 (2017) 1070-1078. DOI https://doi.org/10.1039/C7MH00391A.

[181] C. Suwanpreecha, P. Pandee, U. Patakham, C. Limmaneevichitr, New generation of eutectic Al-Ni casting alloys for elevated temperature services, Mater. Sci. Eng. A 709 (2018) 46-54. https://doi.org/10.1016/j.msea.2017.10.034.

[182] F. Czerwinski, Modern aspects of liquid metal engineering, Metall. Mater. Trans. B 48 (2017) 367-393. https://doi.org/10.1007/s11663-016-0807-6.

[183] D.D. Gu, W. Meiners, K. Wissenbach, R. Poprawe, Laser additive manufacturing of metallic components: materials, processes and mechanisms, Int. Mater. Rev. 57 (2012) 133-164. https://doi.org/10.1179/1743280411Y.0000000014.

[184] P. Agrawal, S. Gupta, S. Thapliyal, S. Shukla, R.S. Haridas, R.S. Mishra, Additively manufactured novel Al-Cu-Sc-Zr alloy: Microstructure and mechanical properties, Addit. Manuf. 37 (2021) 101623. https://doi.org/10.1016/j.addma.2020.101623.

[185] E.O. Olakanmi, R.F. Cochrane, K.W. Dalgarno, A review on selective laser sintering/melting (SLS/SLM) of aluminium alloy powders: Processing, microstructure, and properties, Prog. Mater. Sci. 74 (2015) 401-477. https://doi.org/10.1016/j.pmatsci.2015.03.002.

[186] Q. Shen, X. Kong, X. Chen, Fabrication of bulk Al-Co-Cr-Fe-Ni high-entropy alloy using combined cable wire arc additive manufacturing (CCW-AAM): Microstructure and mechanical properties, J. Mater. Sci. Technol. 74 (2021) 136-142. https://doi.org/10.1016/j.jmst.2020.10.037.

[187] L. Thijs, K. Kempen, J.P. Kruth, J. Van Humbeeck, Fine-structured aluminium products with controllable texture by selective laser melting of pre-alloyed AlSi10Mg powder, Acta Mater. 61 (2013) 1809-1819. https://doi.org/10.1016/j.actamat.2012.11.052.

[188] A. Suzuki, T. Miyasaka, N. Takata, M. Kobashi, M. Kato, Control of microstructural characteristics and mechanical properties of AlSi12 alloy by processing conditions of laser powder bed fusion, Addit. Manuf. 48 (2021) 102383. https://doi.org/10.1016/j.addma.2021.102383.

[189] Y.J. Liu, Z. Liu, Y. Jiang, G.W. Wang, Y. Yang, L.C. Zhang, Gradient in microstructure and mechanical property of selective laser melted AlSi10Mg, J. Alloys Compd. 735 (2018) 1414-1421. https://doi.org/10.1016/j.jallcom.2017.11.020.

[190] X. Liu, Y. Liu, Z. Zhou, Q. Zhan, Enhanced strength and ductility in Al-Zn-Mg-Cu alloys fabricated by laser powder bed fusion using a synergistic grain-refining strategy, J. Mater. Sci. Technol. 124 (2022) 41-52. https://doi.org/10.1016/j.jmst.2021.12.078.

[191] K.B. Hyde, A.F. Norman, P.B. Prangnell, The effect of cooling rate on the morphology of primary Al_3Sc intermetallic particles in Al-Sc alloys, Acta Mater. 49 (2001) 1327-1337. https://doi.org/10.1016/S1359-6454(01)00050-7.

[192] D. Bayoumy, D. Schliephake, S. Dietrich, X.H. Wu, Y.M. Zhu, A.J. Huang, Intensive processing optimization for achieving strong and ductile Al-Mn-Mg-Sc-Zr alloy produced by selective laser melting, Mater. Des. 198 (2021) 109317. https://doi.org/10.1016/j.matdes.2020.109317.

[193] A.B. Spierings, K. Dawson, T. Heeling, P.J. Uggowitzer, R. Schäublin, F. Palm, K. Wegener, Microstructural features of Sc-and Zr-modified Al-Mg alloys processed by selective laser melting, Mater. Des. 115 (2017) 52-63. https://doi.org/10.1016/j.matdes.2016.11.040.

[194] N.E. Uzan, R. Shneck, O. Yeheskel, N. Frage, High-temperature mechanical properties of AlSi10Mg specimens fabricated by additive manufacturing using selective laser melting technologies (AM-SLM), Addit. Manuf. 24 (2018) 257-263. https://doi.org/10.1016/j.addma.2018.09.033.

[195] A. Plotkowski, K. Sisco, S. Bahl, A. Shyam, Y. Yang, L. Allard, P. Nandwana, A.M. Rossy, R.R. Dehoff, Microstructure and properties of a high temperature Al-Ce-Mn alloy produced by additive manufacturing, Acta Mater. 196 (2020) 595-608. https://doi.org/10.1016/j.actamat.2020.07.014.

[196] B. Mehta, S. Bengtsson, D. Riabov, E. Natesan, K. Frisk, J. Ahlström, L. Nyborg, Mechanical properties and microstructural characterisation including high temperature performance of Al-Mn-Cr-Zr based alloys tailored for additive manufacturing, Mater. Des. 244 (2024) 113160. https://doi.org/10.1016/j.matdes.2024.113160.

[197] P. Kürnsteiner, P. Bajaj, A. Gupta, M.B. Wilms, A. Weisheit, X. Li, C. Leinenbach, B. Gault, E.A. Jägle, D. Raabe, Control of thermally stable core-shell nano-precipitates in additively manufactured Al-Sc-Zr alloys, Addit. Manuf. 32 (2020) 100910. https://doi.org/10.1016/j.addma.2019.100910.

[198] J. Iwaszko, K. Kudła, Surface remelting treatment of 7075 aluminum alloy microstructural and technological aspects, Mater. Res. Express 7 (2020) 016523. DOI 10.1088/2053-1591/ab625d.

[199] M.M. Quazi, M.A. Fazal, A.S.M.A. Haseeb, F. Yusof, H.H. Masjuki, A. Arslan, Laser-based surface modifications of aluminum and its alloys, Crit. Rev. Solid State Mater. Sci. 41 (2016) 106-131. https://doi.org/10.1080/10408436.2015.1076716

[200] Y. Chi, G. Gu, H. Yu, C. Chen, Laser surface alloying on aluminum and its alloys: A review, Opt. Lasers Eng. 100 (2018) 23-37. https://doi.org/10.1016/j.optlaseng.2017.07.006.

[201] C. Ma, Q. Zhao, X. Liu, Y. Li, Q. Jiang, Introducing high-density growth twins in aluminum alloys by laser surface remelting via templated nucleation of grains, J. Mater. Sci. Technol. 213 (2025) 315-324. https://doi.org/10.1016/j.jmst.2024.06.036.

[202] M.N. Patel, D. Qiu, G. Wang, M.A. Gibson, A. Prasad, D.H. StJohn, M.A. Easton, Understanding the refinement of grains in laser surface remelted Al-Cu alloys, Scr. Mater. 178 (2020) 447-451. https://doi.org/10.1016/j.scriptamat.2019.12.020.

[203] S.M.A. Bukhari, N. Husnain, F.A. Siddiqui, M.T. Anwar, A.A. Khosa, M. Imran, T.H. Qureshi, R. Ahmad, Effect of laser surface remelting on Microstructure, mechanical properties and tribological properties of metals and alloys: A review, Opt. Laser Technol. 165 (2023) 109588. https://doi.org/10.1016/j.optlastec.2023.109588.

[204] M.A. Alirezaei, R. Raiszadeh, G.R. Khayati, TIG surface alloying of pure aluminium with manganese, Weld. World 67 (2023) 1133-1143. https://doi.org/10.1007/s40194-023-01468-9.

[205] Z. Gong, G.D. Zhang, F. Zheng, Z.G. Wen, M.C. Peng, Performance of different TIG remelting processes in hypereutectic Al-Si cast alloy, Eng. Res. Express 1 (2019) 025027. DOI 10.1088/2631-8695/ab50f1.

[206] N. Bhadauria, S. Pandey, P.M. Pandey, P. Vashishtha, A novel technique for surface modification of aluminium alloy using GTAW, J. Braz. Soc. Mech. Sci. Eng. 45 (2023) 57. https://doi.org/10.1007/s40430-022-04000-0.

[207] A. Ardeshiri, M.H. Sohi, A. Safaei, Surface alloying of A2618 aluminum with silicon and iron by TIG process, Surf. Coat. Technol. 310 (2017) 87-92.

https://doi.org/10.1016/j.surfcoat.2016.12.040.

[208] A. Sharma, R. Jain, P. Agrawal, S. Mukherjee, A. Gumaste, D.F. Davis, R.S. Haridas, R.S. Mishra, Enhanced thermal stability in additive friction stir deposited ODS IN9052 Al alloy, Acta Mater. 279 (2024) 120284. https://doi.org/10.1016/j.actamat.2024.120284.

[209] H. Wang, J. Xu, Y. Kang, M. Tang, Z. Zhang, Study on inhomogeneous characteristics and optimize homogenization treatment parameter for large size DC ingots of Al-Zn-Mg-Cu alloys, J. Alloys Compd. 585 (2014) 19-24. https://doi.org/10.1016/j.jallcom.2013.09.139.

[210] Q. Liu, G. Fan, Z. Tan, Z. Li, D. Zhang, J. Wang, H. Zhang, Precipitation of Al_3Zr by two-step homogenization and its effect on the recrystallization and mechanical property in 2195 Al-Cu-Li alloys, Mater. Sci. Eng. A 821 (2021) 141637.

[211] https://doi.org/10.1016/j.msea.2021.141637. S. Mondol, S.K. Makineni, S. Kumar, K. Chattopadhyay, Enhancement of high temperature strength of 2219 alloys through small additions of Nb and Zr and a novel heat treatment, Metall. Mater. Trans. A 49 (2018) 3047-3057. https://doi.org/10.1007/s11661-018-4614-3.

[212] M. Chemingui, M. Khitouni, K. Jozwiak, G. Mesmacque, A. Kolsi, Characterization of the mechanical properties changes in an Al-Zn-Mg alloy after a two-step ageing treatment at 70 and 135 C, Mater. Des. 31 (2010) 3134-3139. https://doi.org/10.1016/j.matdes.2009.12.033.

[213] O.E. Alarcon, A.M.M. Nazar, W.A. Monteiro, The effect of microstructure on the mechanical behavior and fracture mechanism in a 7050-T76 aluminum alloy, Mater. Sci. Eng. A 138 (1991) 275-285. https://doi.org/10.1016/0921-5093(91)90697-L.

[214] Y.P. Xiao, Q.L. Pan, W.B. Li, X.Y. Liu, Y.B. He, Influence of retrogression and re-aging treatment on corrosion behaviour of an Al-Zn-Mg-Cu alloy, Mater. Des. 32 (2011) 2149-2156. https://doi.org/10.1016/j.matdes.2010.11.036.

[215] S.M. Ashrafizadeh, A.R. Eivani, H.R. Jafarian, J. Zhou, Improvement of mechanical properties of AA6063 aluminum alloy after equal channel angular pressing by applying a two-stage solution treatment, Mater. Sci. Eng. A 687 (2017) 54-62. https://doi.org/10.1016/j.msea.2017.01.024.

[216] Z. Zhang, J. Yu, D. He, Influence of contact solid-solution treatment on microstructures and mechanical properties of 7075 aluminum alloy, Mater. Sci. Eng. A 743 (2019) 500-503. https://doi.org/10.1016/j.msea.2018.11.108.

[217] X. Xu, Y. Zhao, X. Wang, Y. Zhang, Y. Ning, Effect of rapid solid-solution induced by electropulsing on the microstructure and mechanical properties in 7075 Al alloy, Mater. Sci. Eng. A 654 (2016) 278-281. https://doi.org/10.1016/j.msea.2015.12.036.

[218] Y.L. Chang, F.Y. Hung, T.S. Lui, Enhancing the tensile yield strength of A6082 aluminum alloy with rapid heat solutionizing, Mater. Sci. Eng. A 702 (2017) 438-445. https://doi.org/10.1016/j.msea.2017.07.021.

[219] S. Chen, K. Chen, G. Peng, L. Jia, P. Dong, Effect of heat treatment on strength, exfoliation corrosion and electrochemical behavior of 7085 aluminum alloy, Mater. Des. 35 (2012) 93-98. https://doi.org/10.1016/j.matdes.2011.09.033.

[220] T. Ramgopal, P.I. Gouma, G.S. Frankel, Role of grain-boundary precipitates and solute-depleted zone on the intergranular corrosion of aluminum alloy 7150, Corrosion 58 (2002) 687-697. https://doi.org/10.5006/1.3287699.

[221] J.F. Li, N. Birbilis, C.X. Li, Z.Q. Jia, B. Cai, Z.Q. Zheng, Influence of retrogression temperature and time on the mechanical properties and exfoliation corrosion behavior of aluminium alloy AA7150, Mater. Charact. 60 (2009) 1334-1341. https://doi.org/10.1016/j.matchar.2009.06.007.

[222] W. Wang, Q. Pan, X. Wang, Y. Sun, J. Ye, G. Lin, S. Liu, Z. Huang, S. Xiang, X. Wang, Y. Liu, Non-isothermal aging: A heat treatment method that simultaneously improves the mechanical properties and corrosion resistance of ultra-high strength Al-Zn-Mg-Cu alloy, J. Alloys Compd. 845 (2020) 156286. https://doi.org/10.1016/j.jallcom.2020.156286.

[223] D. Dumont, A. Deschamps, Y. Brechet, A model for predicting fracture mode and toughness in 7000 series aluminium alloys, Acta Mater. 52 (2004) 2529-2540. https://doi.org/10.1016/j.actamat.2004.01.044.

[224] M. Zhang, T. Liu, C. He, J. Ding, E. Liu, C. Shi, J. Li, N. Zhao, Evolution of microstructure and properties of Al-Zn-Mg-Cu-Sc-Zr alloy during aging treatment, J. Alloys Compd. 658 (2016) 946-951. https://doi.org/10.1016/j.jallcom.2015.10.296.

[225] Y. Wang, L. Cao, X. Wu, X. Lin, T. Yao, L. Peng, Multi-alloying effect of Ti, Mn, Cr, Zr, Er on the cast Al-Zn-Mg-Cu alloys, Mater. Charact. 201 (2023) 112984. https://doi.org/10.1016/j.matchar.2023.112984.

[226] A.C. Reddy, S.S. Rajan, Influence of ageing, inclusions and voids on ductile fracture mechanism in commercial Al-alloys, Bull. Mater. Sci. 28 (2005) 75-79. https://doi.org/10.1007/BF02711176.

[227] G.T. Motsi, P.A. Olubambi, T.J. Sono, L. Shoke, In situ electron microscopy studies on the tensile deformation mechanisms in aluminium 5083 alloy, Adv. Mater. Res. 1019 (2014) 103-111. https://doi.org/10.4028/www.scientific.net/AMR.1019.103.

[228] R.J. Lane, M.S. Kesler, K. Nawaz, R. Mirzaeifar, Investigating the failure behavior of cast Al-11Ce-0.4 Mg alloys using in-situ scanning electron microscopy tensile testing, J. Alloys Compd. 947 (2023) 169491. https://doi.org/10.1016/j.jallcom.2023.169491.

[229] L.L. Mishnaevsky Jr, N. Lippmann, S. Schmauder, P. Gumbsch, In-situ observation of damage evolution and fracture in AlSi7Mg0.3 cast alloys, Eng. Fract. Mech. 63 (1999) 395-411. https://doi.org/10.1016/S0013-7944(99)00027-2.

[230] J.W. Yeh, W.P. Liu, The cracking mechanism of silicon particles in an A357 aluminum alloy, Metall. Mater. Trans. A 27 (1996) 3558-3568. https://doi.org/10.1007/BF02595447.

[231] M. Besterci, O. Velgosová, J. Ivan, P. Hvizdoš, I. Kohútek, Influence of volume fraction on fracture mechanism of Al-Al_4C_3 system studied by "in-situ tensile test in SEM", Kovove Mater. 46 (2008) 139-143.

[232] H.C. Fang, F.H. Luo, K.H. Chen, Effect of intermetallic phases and recrystallization on the corrosion and fracture behavior of an Al-Zn-Mg-Cu-Zr-Yb-Cr alloy, Mater. Sci. Eng. A 684 (2017) 480-490. https://doi.org/10.1016/j.msea.2016.12.009.

[233] D. He, Y. Chen, S. Chen, Y. Lin, J. Wu, High-temperature tensile characteristics of an Al-Zn-Mg-Cu Alloy: fracture characteristics and a physical mechanism constitutive model, Materials 17 (2024) 2628. https://doi.org/10.3390/ma17112628.

[234] J.H. Sokolowski, M.B. Djurdjevic, C.A. Kierkus, D.O. Northwood, Improvement of 319 aluminum alloy casting durability by high temperature solution treatment, J. Mater. Process. Technol. 109 (2001) 174-180. https://doi.org/10.1016/S0924-0136(00)00793-7.

[235] R. Singer, W. Blum, B. Ilschner, Deformation-induced microstructural instability in a θ′-hardened aluminum alloy at high temperature, Mater. Sci. Eng. 40 (1979) 235-243. https://doi.org/10.1016/0025-5416(79)90194-0.

[236] B.K. Milligan, S. Roy, C.S. Hawkins, L.F. Allard, A. Shyam, Impact of microstructural stability on the creep behavior of cast Al–Cu alloys, Mater. Sci. Eng. A 772 (2020) 138697.https://doi.org/10.1016/j.msea.2019.138697.

[237] Y.H. Gao, L.F. Cao, J. Kuang, J.Y. Zhang, G. Liu, J. Sun, Assembling dual precipitates to improve high-temperature resistance of multi-microalloyed Al-Cu alloys, J. Alloys Compd. 822 (2020) 153629. https://doi.org/10.1016/j.jallcom.2019.153629.

[238] D. Li, K. Liu, J. Rakhmonov, X.G. Chen, Enhanced thermal stability of precipitates and elevated-temperature properties via microalloying with transition metals (Zr, V and Sc) in Al-Cu 224 cast alloys, Mater. Sci. Eng. A 827 (2021) 142090. https://doi.org/10.1016/j.msea.2021.142090.

[239] S. Bahl, J.U. Rakhmonov, C. Kenel, D.C. Dunand, A. Shyam, Effect of grain-boundary θ-Al2Cu precipitates on tensile and compressive creep properties of cast Al-Cu-Mn-Zr alloys, Mater. Sci. Eng. A 840 (2022) 142946. https://doi.org/10.1016/j.msea.2022.142946.

Chapter 3

Experimental Instruments and Characterization Techniques

This section provides an overview of the experimental methods and analyses employed in this thesis. It will first explain the alloying of Al7075 (Al-Zn-Mg-Cu) and Al2219 (Al-Cu) systems by incorporating multiple solutes such as Ni, Mn, Ti, Sc, and Zr using a vacuum resistance melting furnace. The rapid cooling rates were achieved using a copper mould gravity die casting technique to produce the cast alloys. The cast alloys were heat treated to enhance mechanical properties through precipitation hardening, and re-aging techniques were subsequently applied to understand the microstructural stability at 300°C. Test specimens were precisely fabricated using wire-electrical discharge machining (W-EDM) for flat sub-size tensile and creep testing, while cylindrical specimens for high-temperature testing were machined using CNC machine. The microstructural characterization was performed at each processing stage, from as-cast to re-aging conditions. The copper mould casting technique was specifically chosen to enhance solute solubility and potentially generate metastable phases. The composition of the developed alloys was analysed using the Optical Emission Spectrometer (OES). JMatPro software was used to perform thermodynamic simulations, and Differential Scanning Calorimetry (DSC) was carried out to evaluate the phase stability and precipitation behavior. These microstructural and phase analyses were performed using optical and scanning electron microscopy (SEM), transmission electron microscopy (TEM), electron backscattered diffraction (EBSD), atom probe tomography (APT) and X-ray diffraction (XRD) techniques. The properties of the material were evaluated through Brinell hardness, Vickers microhardness, ambient and elevated temperature tensile testing, and tensile creep testing. Additionally, real-time observation of microscale deformation and fracture mechanisms was carried out using an in-situ tensile testing setup within a field emission scanning electron microscope (FESEM), which also contributed to the optimization of microalloying elements (Sc, Zr, Ti) in the Al7075 alloy. Figure 3.1 summarizes the experimental methodology, alloy composition variation and characterization techniques used in this study. The developed material is expected to exhibit enhanced yield strength and plastic deformation at both ambient and elevated temperatures compared to conventional Al-Zn-Mg-Cu and Al-Cu alloys. To further refine microstructure and properties, optimized alloy compositions were subjected to surface re-melting via TIG-assisted arc processing. This integrated approach aimed to develop high-performance aluminium alloys with improved strength at ambient and high temperature.

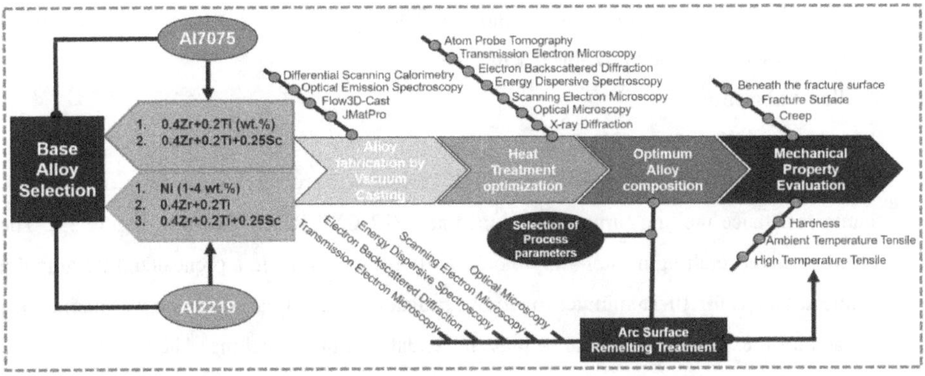

Figure 3.1 A detailed flow diagram summarizing the experimental methodology and characterization techniques used in this study.

3.1 Base alloy selection and compositional variations

In this study, Al7075, a high-strength aluminum alloy, and Al2219, an alloy recognized for its excellent weldability and moderate high-temperature strength, were chosen as base materials due to their potential for further enhancement through alloying. For Al7075, two compositional modifications were examined: (1) the addition of 0.4 wt.% Zr and 0.2 wt.% Ti and; (2) the combined addition of 0.4 wt.% Zr, 0.2 wt.% Ti, and 0.25 wt.% Sc to refine its grain structure, improve strength and thermal stability by the formation of $L1_2Al_3TM$ dispersoids. The alloys were prepared using 99.99% pure Al, 99.99% pure Mg, Al-30Cu, Zn-5Al, Al-10Zr, Al-10.22Ti and Al-2Sc master alloys. These master alloys were selected to precisely control the composition and ensure uniform distribution of alloying elements during melting. The detailed compositions of the Al7075 base alloy and microalloying additions are provided in Chapter 4 (Table 4.1) and Chapter 5 (Table 5.1).

The high-temperature properties of Al2219 were addressed through three tailored alloying approaches: (1) the addition of 1 wt.%-4 wt.% Ni to improve thermal stability and high-temperature strength and creep resistance, (2) the inclusion of 0.4 wt.% Zr and 0.2wt.% Ti and (3) the combination of 0.4 wt.% Zr, 0.2wt.% Ti, and 0.25 wt.% Sc to significantly enhance strength and resistance to thermal degradation. These compositional variations were designed to improve the material strength under high-temperature applications. The alloys were prepared using 99.9% pure Al and 99.5% pure master alloys, including Al-35Cu, Al-10Mn, Al-10Ni, Al-10Zr, Al-10.22Ti and Al-2Sc. All Master alloys were supplied by Ceraflux India Pvt. Ltd., and their chemical compositions were provided by the supplier and independently verified using a Metavision 1008i Optical Emission Spectrometer (OES). The chemical compositions of all master alloys are provided in Appendix A. The detailed compositions of these modified Al2219 alloys are provided in Table 6.1 in Chapter 6.

3.2 Alloy Fabrication

3.2.1 Vacuum Resistance Melting

The base and microalloyed Al7075 and Al2219 alloys were fabricated using an in-house vacuum resistance melting furnace developed at CSIR-CMERI, as shown in Figure 3.2. The process involved melting master alloy blocks and pure materials in a preheated 3 kg graphite crucible at 500°C for 10-15 minutes to remove moisture. The furnace chamber was maintained at a vacuum pressure of 10^{-3} mbar to prevent oxidation during melting. The molten material was stirred for 10 minutes to ensure compositional homogeneity, followed by argon gas purging at a constant flow rate of 10 l/min for 10 minutes to remove impurities and minimize microporosity. After degassing, slag was removed, and the molten material was poured into a rectangular copper mold with dimensions of 148 mm × 60 mm × 46 mm, as shown in Figure 3.3. The mold was immersed in chilled water to achieve a high cooling rate of ~15°C/sec. The casting chamber was equipped with a thermocouple and a digital temperature controller to ensure precise temperature monitoring throughout the process. Radiographic analysis was conducted to detect internal defects such as porosity or cracks, ensuring the cast blocks were of high quality, defect free and suitable for mechanical and microstructural evaluations. The specific process parameters for fabricating each composition are described in detail in the respective chapters.

Figure 3.2 Vacuum resistance melting furnace setup.

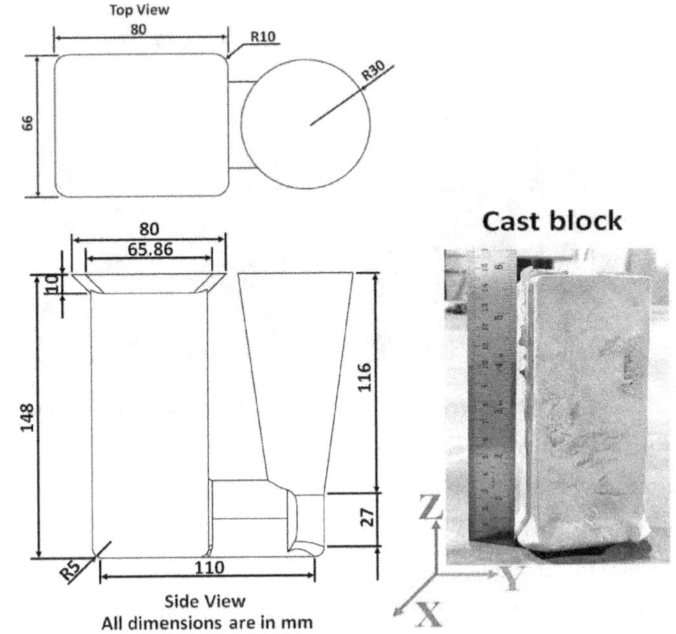

Figure 3.3 Copper permanent mould used for casting ingots (all dimensions are in mm).

3.2.2 Heat Treatment

A muffle furnace with a maximum operating temperature of 1400°C and a hot air oven capable of reaching 300°C were utilized for the solutionizing and aging treatments, respectively. The muffle furnace (Figure 3.4(a)) was calibrated according to manufacturer specifications using built-in thermocouples. An external thermocouple was employed to verify temperature uniformity, ensuring deviations remained within ±5°C of the setpoint. During solutionizing, a controlled ramp rate of 6°C/min was maintained.

The aging process was conducted in the hot air oven (Figure 3.4(b)) which underwent similar calibration to ensure precise temperature regulation. Optimal solutionizing and aging parameters, including temperature and time, were established for the Al-Zn-Mg-Cu alloy system, as detailed in Chapters 4, and 5 and the Al-Cu-Mn alloy system, in Chapters 6 and 7. These parameters were derived through systematic experimentation. Full heat treatment procedures, including alloy-specific conditions, are documented in the experimental procedure sections of the respective chapters.

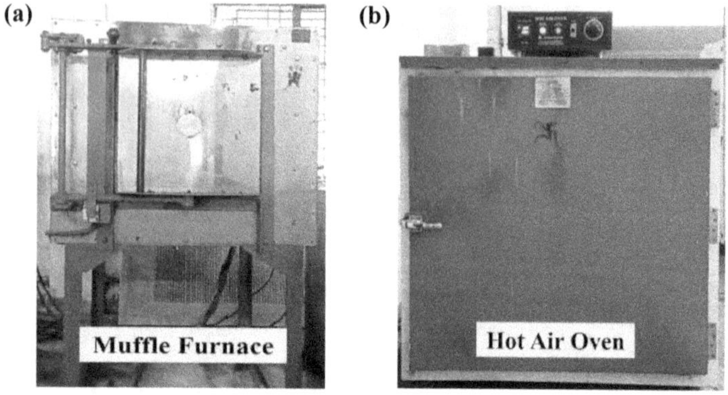

Figure 3.4 Heat treatment furnaces (a) muffle furnace and (b) hot air oven.

3.2.3 Arc Surface re-melting

The optimized alloy composition underwent arc surface remelting (ASR) treatment to enhance its mechanical properties and refine its microstructure. Figure 3.5 illustrates the experimental setup for arc surface remelting (ASR). The process utilizes a TIG welding torch connected to a welding power source (Fronius iWave 300i AC/DC TIG), which generates the arc required for remelting. The TIG torch is mounted on an ABB IRB 1410 robotic manipulator, controlled by an IRC5 robotic controller. The schematic highlights the unidirectional scanning trajectory, with the remelted zones on both surfaces of the cast plate visually demarcated by red dashed lines. Prior to remelting, the cast block was cleaned with acetone to remove the grease or oil before the remelting. The details of the remelting parameters can be found in the experimental procedure section of Chapter 8.

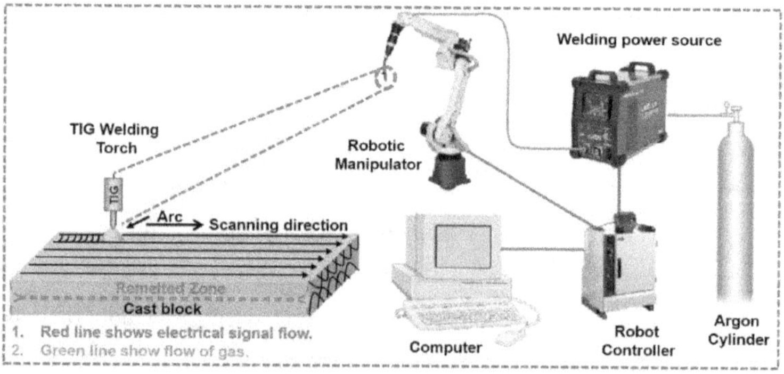

Figure 3.5 Schematic diagram of the TIG assisted arc surface remelting treatment of the extracted cast plate.

3.3 Mechanical Testing

3.3.1 Hardness Test

The bulk hardness of the developed alloys was measured using a Brinell hardness tester according to the ASTM E10 standard, as shown in Figure 3.6a (Bluestar B-3000 TS), employing a 10 mm ball under a 1000 kg load. The phase hardness of as-cast and heat-treated conditions, along with different re-melted regions of the arc surface re-melted (ASR) specimens was evaluated using Vickers microhardness testing with a Falcon 500 tester (Innova Test, Netherlands), as illustrated in Figure 3.6b. This involved a pyramidal type of diamond indenter having 120° tip angle was used for the analysis with 4.9 N load and 15 s dwell time.

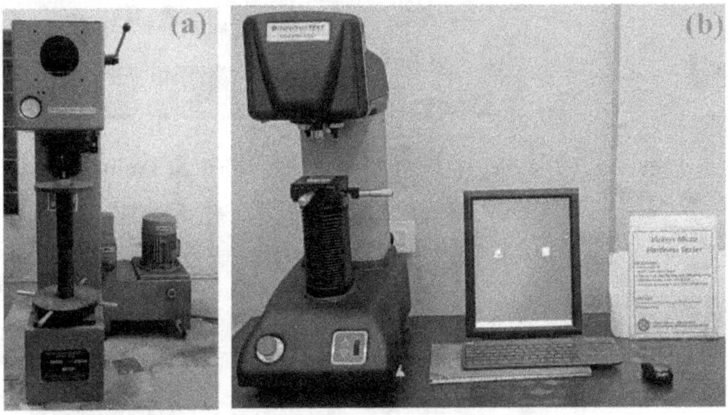

Figure 3.6 (a) Bluestar B-3000 TS Brinell hardness tester (b) Falcon 500, (Innova Test, Netherland) Vicker's microhardness tester.

3.3.2 Monotonic Tensile Test at ambient and elevated temperatures

The ambient temperature tensile tests were conducted using flat dog-bone-type samples. The tensile tests were carried out at a constant crosshead displacement rate of 0.5 mm/min by an UTM (H50KS, Tinius Olsen, UK) tensile test machine with a maximum load capacity of 50 kN (Figure 3.7).

The elevated temperature tensile tests were conducted on round specimens as per ASTM E8M standards using UTM (H100KGS, BLISS, ITW, USA). The tests were performed at temperatures up to 350°C and at a constant initial strain rate of 5.0×10^{-4} s^{-1}. Figure 3.8 illustrates the setup for elevated temperature tensile testing. An electric resistance heating furnace was integrated into the system to maintain the desired test temperatures, and a thermocouple was attached to monitor the specimen temperature accurately. Prior to testing, the specimens were held at the respective test temperature for 45 minutes to ensure uniform thermal stabilization throughout the gauge length. A linear variable differential transformer

(LVDT) extensometer (10 μm resolution) was mounted on the specimen shoulders. The yield stress was calculated using the 0.2% offset and elongation was measured at fracture. The details of the experiments were described in chapter 5 and 7.

Figure 3.7 Tensile testing using UTM (H50KS, Tinius Olsen, UK).

Figure 3.8 Ambient and elevated temperature tensile testing using a servo hydraulic UTM (H100KGS, BISS, ITW, India).

3.3.3 In-situ Tensile testing under FESEM

In-situ tensile testing was conducted using a Deben Microtest 2-kN testing stage positioned within vaccum chamber of an FE-SEM (Sigma HD, Carl Zeiss, Germany). The experimental setup is illustrated in Figure 3.9. The specimens were horizontally mounted and securely clamped between two jaws using a small-lead, precise ball screw, then supported on stainless steel slide bearings. Symmetrical movement of the jaws in opposite directions was achieved through a dual-threaded lead screw, ensuring the sample remained centered in the field of view.

The microtest controller (Figure 3.9b), regulating a standard speed range of 0.033-0.4 mm/min, maintained precision through an optical encoder-fitted motor with an accuracy level of approximately 1%. Figure 3.9c depicts the tensile loading direction of the specimen in the microtester. During tensile testing, the sample displacement speed was set at 0.033 mm/min. The graphical user interface (GUI) affords the user comprehensive control over every aspect of the tensile testing machine, as shown in Figure 3.9d. SEM images were acquired at different loading intervals.

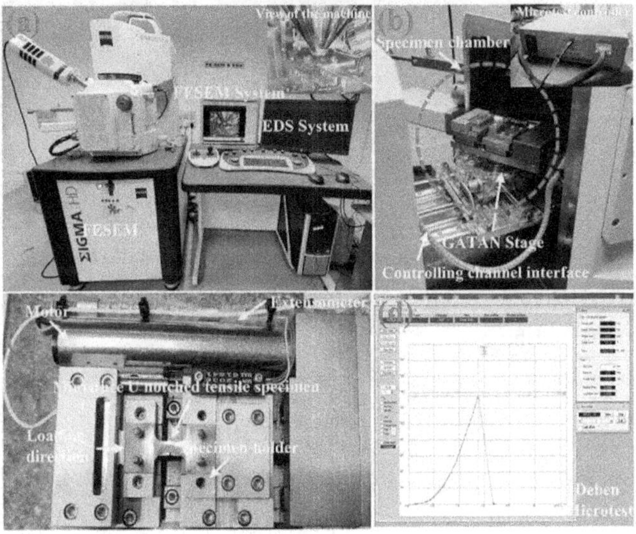

Figure 3.9 In-situ experimental set-up of (a) FESEM (Sigma HD, Carl Zeiss, Germany) (b) FESEM specimen chamber with microtest controller (c) enlarged view of the microtester loading direction with U notched miniature tensile specimen (d) The user interface includes real-time plot of the applied load vs elongation.

3.3.4 Tensile Creep Testing

The tensile creep tests for the final optimized alloy were conducted using the STS-9018 single-lever arm creep testing machine (as shown in the Figure 3.10) with a 30 kN load capacity. The tensile creep specimens were fabricated following the ASTM E139 standard. The machine features a high-temperature electric resistance furnace with a K-type thermocouple (Ni/Al-Ni/Cr) attached at the gauge length for accurate temperature monitoring. Creep strain was measured using a linear variable differential transformer (LVDT) extensometer (10 μm resolution) mounted on the specimen shoulders. Tests were performed at 250-300°C under different stress levels of 20-80 MPa to evaluate high-temperature creep resistance. The details of the experiments were described in chapter 7.

Figure 3.10 STS-9018 single-lever arm type tensile creep testing machine.

3.4 Characterization Techniques

3.4.1 Sample Polishing

High-quality microstructural analysis was achieved by meticulously preparing the samples using the following systematic polishing procedure.

- ✓ Small sections of samples were cut from cast, heat-treated, and arc surface remelted blocks using a computer-controlled wire electrical discharge machine (W-EDM). The sample sectioned position and size will be illustrated in each chapter.
- ✓ Due to the small size of the samples, they were mounted on metallic or polymer stubs for easier handling.
- ✓ The mounted samples were then ground and polished by a Bainpol Metco, India, grinder-polisher machine shown in Figure 3.11. Initial polishing was performed using different grades of emery paper on a rotating wheel at a controlled speed of 150 rpm, with water as a suspension.
- ✓ The sequence of emery papers used included grades 400, 600, 800, 1000, 1200, 1500, 2000, and 2500. The polishing direction was changed by 90° with each paper.
- ✓ Polishing on each emery paper continued in the same direction until all scratches from the previous paper were removed.

- ✓ After the final emery paper polishing, cloth polishing was performed using a diamond paste of 3-4 μm and 0.5-1.0 μm particle size with a colloidal solution and water until a mirror finish was achieved.

Finally, samples were etched using Keller's reagent (composed of 2.5% HNO_3, 1.5% HCl, and 1% HF in water) for 10 seconds (etching time depends on alloy composition) to reveal the microstructure. This process was essential for enhancing the visibility of grain boundaries, secondary phases, and other microstructural features, enabling detailed analysis through optical microscopy (OM), Scanning Electron Microscopy (SEM) and electron backscatter diffraction (EBSD) techniques.

Figure 3.11 Bainpol Metco grinding-polishing machine.

3.4.2 X-ray Diffraction (XRD)

X-ray diffraction (XRD) with a Cu Kα radiation source (λ=1.5418Å) was performed to identify the phase constituents by a Regaku Minifex (Figure 3.12) of as-cast, solution treated and aged specimens. The accelerating voltage and tube current were set at 35 kV and 28.6 mA, respectively. The crystallite size, dislocation density, and microstrain were quantitatively determined through detailed analysis of X-ray diffraction (XRD) peak profiles. The methodology for these calculations-including peak broadening analysis, Williamson-Hall plot for microstrain separation, and dislocation density is comprehensively described in the Materials and Methods section of each respective chapter.

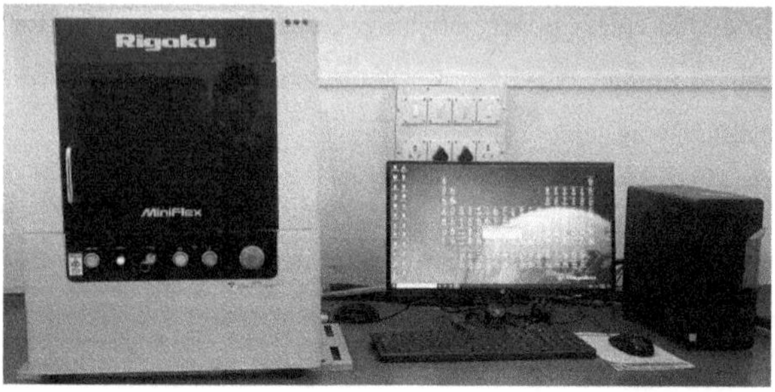

Figure 3.12 Regaku Minifex X-ray diffractometer.

3.4.3 Differential Scanning Calorimetry (DSC)

Thermal analysis was performed using a NETZSCH STA 449 F1 Jupiter differential scanning calorimeter (DSC) (Figure 3.13) to investigate the dissolution temperature of precipitates, melting characteristics, and solutionizing temperature of the developed alloys. The experiment was carried out in an inert argon atmosphere to avoid oxidation. A sample weight of ~25 mg was utilized, and the DSC heating cycles were executed within a temperature range of 25°C to 800°C at a constant heating rate of 10 °C/min. This allowed for the analysis of enthalpy and free energy changes in both as-cast and heat-treated samples.

Figure 3.13 NETZSCH STA 449 F1 Jupiter differential scanning calorimeter.

3.4.4 Optical Microscopy (OM)

The overall microstructures of the etched samples, including dendrites, precipitations, intermetallic and dispersoids of as-cast and heat-treated samples were captured using Axio-ImagerA1m, ZEISS optical microscope, as shown in Figure 3.14. Images were obtained using direct bright field imaging at different magnifications.

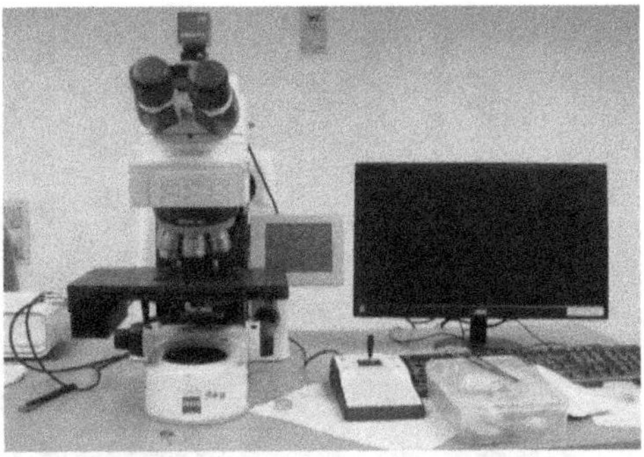

Figure 3.14 Axio-ImagerA1m, ZEISS optical microscope.

3.4.5 Field Emission Scanning Electron Microscopy (FE-SEM)

The detailed observation of the matrix microstructure, precipitation, fracture mode, beneath the fracture surfaces and chemical composition was preliminarily performed using Axio-ImagerA1m, ZEISS, Oberkochen, Germany and JEOL JSM-7200f Scanning Electron Microscope (SEM) equipped with Oxford Instruments X-Max Energy Dispersive X-ray Spectroscopy (EDS) detector, as shown in Figure 3.15 and Figure 3.16 respectively, operating at 20 kV and working distance at 6 mm.

Backscattered scanning electron (BSE) imaging conducted using Axio-ImagerA1m, ZEISS FESEM and Electron Backscattered Electron Diffraction (EBSD) was conducted using JEOL JSM-7200f SEM equipped with Oxford Instruments Nordlys II(S) high resolution EBSD camera. AZTech software integrated with SEM machine used for data acquisition and AZCrystal was used for data processing and analysis. The EBSD maps were collected at 20 kV accelerating voltage, 21 mm working distance, and 1 μm step size.

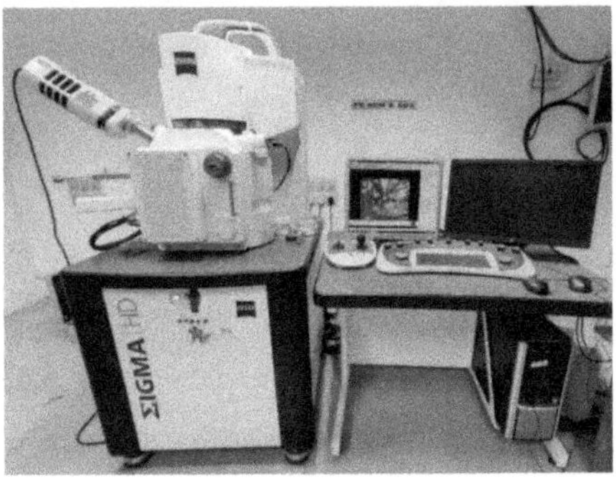

Figure 3.15 Carl Zeiss ∑igma HD field emission scanning electron microscope (FESEM) equipped with energy-dispersive X-ray spectroscopy (EDS).

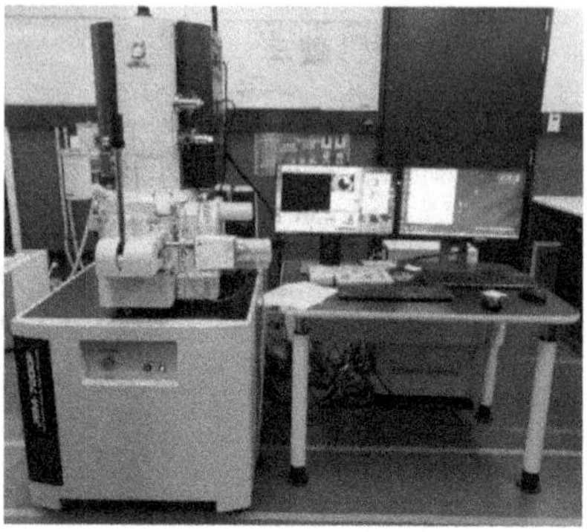

Figure 3.16 JEOL JSM-7001F Scanning Electron Microscope.

3.4.6 Transmission electron microscopy and allied analysis

Thin TEM foils were initially thinned down to 100 μm using mechanical polishing, followed by punching 3 mm discs from as-cast, heat-treated, and arc surface remelted specimens. These discs were further mechanically ground and dimpled to a thickness of 50 μm to prevent shadowing in the central region. Final thinning to perforation was achieved through twin-jet electro-polishing using (Struers-Tenupol unit) an electrolyte containing 30% nitric acid in 70% methanol, operated at 13 V and -30°C.

Detailed transmission electron microscopy (TEM) investigations were performed on a Tecnai 20T G2 TEM (Figure 3.17) at 200 kV and Spectra 300 aberration corrected HR(S)TEM with EDS and EELS operated at 300 kV. Both conventional TEM and STEM imaging techniques were utilized. The α-Al matrix, precipitates and dispersoids were imaged using centred dark-field imaging, either by selecting a low-index superlattice reflection of the phase or indirectly through strain-field contrast using a fundamental reflection. The elemental mapping of precipitates and dispersoids was performed in STEM mode utilizing an EDS detector.

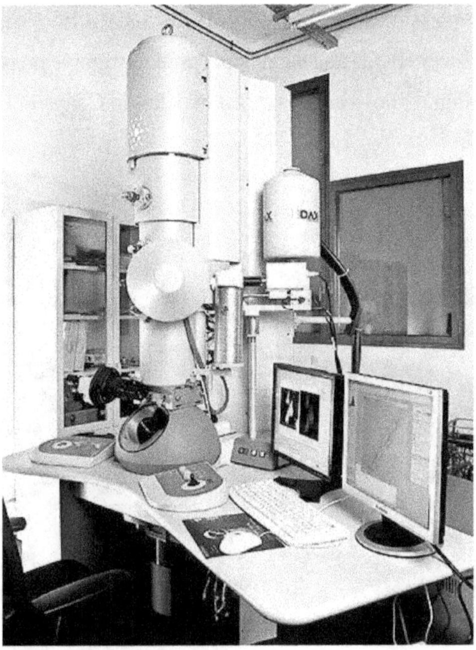

Figure 3.17 Tecnai 20T G2 Transmission electron Microscope (TEM).

3.4.7 3D Atom Probe Tomography (APT)

The three-dimensional distribution of elements at the atomic scale was quantitatively analyzed using atom probe tomography (APT) with a local electrode atom probe (LEAP™ 5000XR, Cameca Instruments, USA). The LEAP 5000 XR (Figure 3.18) is a state-of-the-art instrument designed for high-resolution compositional mapping of materials at the atomic scale. It employs a reflection ion optic design and operates in both voltage and laser modes, achieving a mass resolving power of 1100 and a field of view up to 200 nm. The system enables analysis at cryogenic temperatures, ranging from 25K to 80K, and utilizes a 355 nm UV laser with an energy range from 0.001 pJ to 1000 pJ, supporting a pulse rate of up to 500 kHz.

The sample preparation process (Figure 3.19) involves fabricating sharp needle-shaped specimens using Focused Ion Beam (FIB) milling techniques. The site-specific lift-out of APT tips was performed using a dual-beam electron microscope equipped with a focused ion beam (Thermo Fisher Scientific, Helios G4 UX) (Figure 3.20). It includes advanced detectors such as in-lens TLD, secondary electron detectors, and in-column detectors, along with application software like AutoFIB and AutoTEM 4. Together, these systems enabled high-resolution imaging and analysis for APT studies. The data was then reconstructed and analyzed using the

Integrated Visualization and Analysis Software (IVAS 3.8.10) from Cameca Instruments Inc. (Madison, WI, USA). The methods and parameters used for sample preparation in APT analysis of the developed aluminium alloys are described in detail in Chapter 6.

Figure 3.18: LEAP 5000XR atom probe tomography (APT).

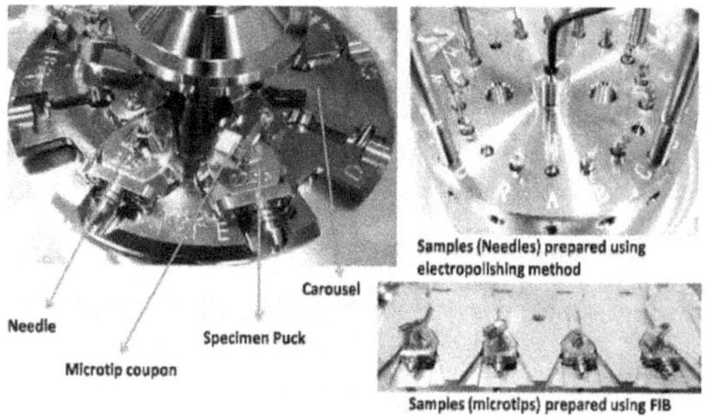

Figure 3.19: Sample preparation setup for APT, showing microtips fabricated using FIB.

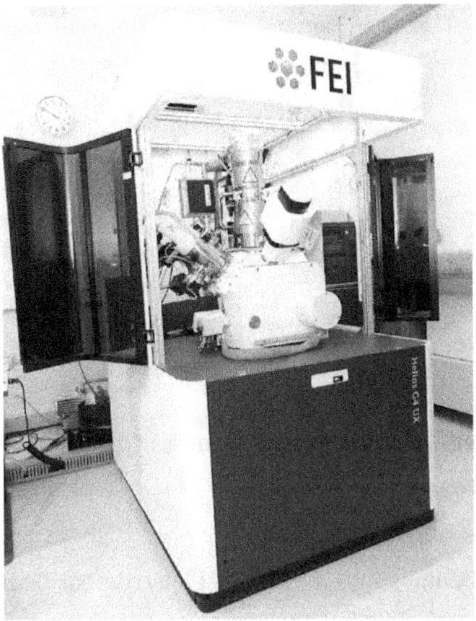

Figure 3.20 Helios G4 UX FIB electron microscope.

3.5 Analytical characterization

3.5.1 Optical emission spectroscopy (OES)

The composition of the developed aluminium alloys was analysed using the Metavision 1008i Optical Emission Spectrometer (OES) (Figure 3.21) to ensure accuracy and consistency in elemental distribution. The analysis was conducted on cylindrical samples with a diameter of 12 mm and a height of 8 mm, prepared using Wire Electrical Discharge machining (EDM). A high-energy spark was generated between the electrode and the sample surface, vaporizing a small portion of the material, and the emitted light spectrum was analysed to determine the concentration of alloying elements such as Cu, Zn, Mg, Mn, Ni, Ti, Zr and Sc. To ensure representative composition data, a minimum of 15 sparks were collected from different locations on each sample. The analysis was conducted under a controlled argon gas flow rate of 1.0-1.5 L/min to maintain a stable plasma and minimize contamination. The detailed composition of each alloy is systematically listed in the respective chapters of the thesis, corresponding to their specific experimental investigations and analyses.

Fig. 3. 21 Metavision 1008i (Metavision, India) Optical Emission Spectrometer (OES).

3.5.2 JMatPro software

JMatPro software was employed to predict the phase stability and transformation behaviour of the developed aluminium alloys based on their compositions. JMatPro, a computational thermodynamic and materials property simulation tool, utilizes CALPHAD-based models to generate phase diagrams, solidification behaviour, and equilibrium phase fractions over a range of temperatures. These computational predictions provided crucial insights into the alloy design, ensuring an optimized microstructure tailored for material property.

3.5.3 FLOW-3D® and ANSYS™ software

The solidification behaviour of the developed Al-Zn-Mg-Cu alloys was simulated using FLOW-3D® software. The computational model incorporated material properties, boundary conditions, and heat transfer to predict the temperature evolution and cooling rates during solidification in a copper mould, as detailed in Chapter 5.

Finite element analysis (FEA) of the in-situ tensile specimen was conducted using ANSYS™ Workbench (version 19.2) to understand the deformation behaviour and stress distribution on the gauge length of the developed Al-Zn-Mg-Cu-X alloys. The simulation employed a meshed geometry of the tensile specimen, incorporating material properties and boundary conditions representative of the experimental setup. This numerical approach enabled a deeper understanding of strain localization, stress evolution, and potential failure mechanisms in the studied alloys. These details are thoroughly discussed in Chapter 4.

3.5.4 Image J software

The microstructural features of Al-Zn-Mg-Cu and Al-Cu-Mn-Ni alloys in both as-cast, heat-treated and arc-surface remelted conditions were quantitatively analysed using ImageJ software. The analysis included measurements of α-Al cell size, eutectic volume fraction, and intermetallic particle size and distribution based on OM and SEM images. The length and thickness of θ', η', T, and S phases, as well as the size and distribution of Al_3TM core-shell dispersoids, were determined from TEM images.

3.6.4.1 α-Al cell Size estimation

The microstructural refinement of the α-Al phase was evaluated using the standardized line intercept method, implemented through ImageJ's measurement tools. Five equidistant horizontal and five vertical measurement lines were systematically overlaid on each micrograph, covering the entire field of view to obtain representative data. The number of intersections between these measurement lines and eutectic phase boundaries was manually counted to determine the linear intercept density (P_L). The mean α-Al cell size was then calculated as the reciprocal of P_L, with the equation:

$$\text{Line intercept } (P_L) = \frac{Number\ of\ intersections}{length\ of\ line}$$

$$\text{Cell size} = \frac{1}{P_L}$$

This approach provided a robust and repeatable method for quantifying the α-Al cell structures, with multiple measurements performed to establish statistical significance.

3.6.4.2 Eutectic volume fraction and intermetallic particle size analysis

The quantitative assessment of eutectic phase distribution was performed through systematic image analysis of optical microscopy (OM) and scanning electron microscopy (SEM) micrographs. The relative proportions of each microstructural constituent, the eutectic region in the microstructure was distinguished. The microstructure images were scanned at a resolution of 1200 dpi. These images were then processed and thresholded using the ImageJ software to estimate the eutectic and α-Al volume fractions. For accuracy, at least five microstructure images were analyzed for each condition in as-cast, heat-treated and arc-surface remelted specimens.

The size distribution of intermetallic phases was analyzed using high-resolution SEM imaging at 40000x magnification. The "Freehand Selection" tool in ImageJ was used to measure the dimensions of at least 100 selected particles per condition. The data was processed to determine the mean size, size distribution, and aspect ratio, offering detailed insights into intermetallic phase morphology.

3.6.4.3 Precipitate and dispersoid size and distribution analysis

Nanoscale precipitate analysis was conducted using transmission electron microscopy (TEM) to resolve fine structural features. For θ' and η' strengthening precipitates, the "Straight Line" measurement tool in ImageJ was employed to determine both length and thickness from bright-field TEM images. Similarly, the complex T-$Al_{20}Cu_2Mn_3$ phases were characterized through careful dimensional analysis. The core-shell Al_3TM dispersoids were evaluated using the "Fit Ellipse" function, which provided measurements of both the core diameter and overall particle size. A minimum of 100 particles per condition was analyzed to ensure statistical significance, with measurements taken from multiple TEM images to account for local variations in precipitate distribution. The volume fraction and number density were calculated using the number of precipitates per unit area with the ImageJ software.

3.6.5 Digital Micrograph software

The DigitalMicrograph® software (Gatan, Inc.) was utilized for the acquisition, processing, and analysis of Transmission Electron Microscopy (TEM) and Scanning Transmission Electron Microscopy (STEM) images. This software facilitated high-resolution imaging, enabling detailed observation of microstructural features, including precipitates, dislocations, and grain boundaries. Some of the Selected Area Electron Diffraction (SAED) patterns were processed using Digital Micrograph for precise phase identification and crystallographic analysis. The advanced filtering and image enhancement tools improved contrast and resolution of the captured microstructure.

Chapter 4

Unravelling crack tip damage mechanisms: In-situ tensile assessment of Al-6Zn-2.1Mg-2Cu alloy strengthened by Ti, Zr, and Sc micro-alloying

Preface

This chapter presents a comprehensive investigation into the impact of precipitates, intermetallic, and dispersoids on the microstructure and crack-tip damage mechanisms of base and micro-alloyed Al-6Zn-2.1Mg-2Cu alloys, as observed during in-situ SEM. With a particular emphasis on the role of micro-alloying elements such as Ti, Zr, and Sc, the research was motivated by the need to enhance the mechanical performance and fracture resistance of high-strength Al alloys. This chapter addresses the first research question by observing real-time microscale deformation, revealing the influence of various micro-alloying strategies that influence crack initiation, propagation, and overall alloy toughness. Four distinct alloy compositions were systematically studied: Alloy A, the base alloy Al-6Zn-2.1Mg-2Cu without microalloying additions; Alloy B, Al-6Zn-2.1Mg-2Cu-0.4Zr-0.2Ti, containing 0.4 wt.% Zr and 0.2 wt.% Ti; Alloy C, Al-6Zn-2.1Mg-2Cu-0.6Zr-0.4Ti, with increased microalloying of 0.6 wt.% Zr and 0.4 wt.% Ti; and Alloy D, Al-6Zn-2.1Mg-2Cu-0.4Zr-0.2Ti-0.25Sc, which incorporates 0.4 wt.% Zr, 0.2 wt.% Ti, and 0.25 wt.% Sc. Optimized micro-alloying additions of Ti, Zr, and Sc form fine, coherent Al_3(Sc, Zr, Ti) dispersoids that arrest cracks and enhance fracture resistance, while excessive micro-alloying leads to coarse, brittle Al_3(Zr, Ti) intermetallic that cause premature failure.

This chapter includes a research article entitled 'Unravelling crack tip damage mechanisms: In-situ tensile assessment of Al-6Zn-2.1Mg-2Cu alloy strengthened by Ti, Zr, and Sc micro-alloying' and published in the journal "Engineering Fracture Mechanics". The exact word form of the published articles has been reproduced in this chapter and the published form has been attached in the last section of this thesis.

Key findings of Chapter 4 were published in the research article as outlined below.

1. **Diya Mukherjee**, Yoshit Tiwari, E Chandrasekar, Manidipto Mukherjee, Nilrudra Mandal, Dong Qiu, Mark A. Easton, Himadri Roy, Unravelling crack tip damage mechanisms: In-situ tensile assessment of Al-6Zn-2Mg-1.8Cu alloy strengthened by Ti, Zr, and Sc micro-alloying, Engineering Fracture Mechanics, 312 (2024), 110663. https://doi.org/10.1016/j.engfracmech.2024.110663.

Abstract

This study investigates the effect of micro-alloying elements (Ti, Zr and Sc) on the crack tip damage behaviour of cast Al-6Zn-2.1Mg-2Cu-X alloys to understand the effect of grain boundary interfaces, matrix, intermetallics and dispersoids on deformation and fracture behaviour. Notched tensile specimens were prepared and subjected to in-situ tensile deformation under FESEM to investigate the crack initiation, propagation, and eventual failure processes under uniaxial tensile loading. The experimental results reveal that fracture in the four Al-6Zn-2.1Mg-2Cu-X alloys is quasi-brittle due to strain accumulation and cleavage fracture mechanisms. The crack mainly initiates from the grain boundary lath S-Al_2CuMg and η-$MgZn_2$ intermetallic phases. Adding 0.25% Sc forms small coherent Al_3(Sc, Zr) dispersoids, increasing crack resistance. The ductility of the matrix is due to void nucleation, crack meandering, and crack tip-blunting phenomena. At high additions of microalloying elements (1 wt.%), large Al_3(Zr, Ti) dispersoids form and agglomerate in alloy C, contributing to fracture through intermetallic cracking. These findings from materials characterization and in-situ tensile testing indicate potential methods to enhance the fracture toughness of Al-6Zn-2.1Mg-2Cu alloys with additions of Zr, Ti, and Sc for future development.

Keywords: Aluminium alloys; Scandium; Small scale specimen testing; In-situ SEM tensile testing; Crack tip damage mechanism.

4.1 Introduction

High-strength cast Al-Zn-Mg-Cu alloys have attracted widespread attention in the aerospace and automotive industries due to their exceptional strength, low density and good fatigue resistance, ensuring a high level of safety [1-4]. Nevertheless, these alloys exhibit exceptional strength but are also susceptible to fracture and stress corrosion cracking [5,6]. In practical applications, the microstructure comprises various heterogeneities that can contribute to localized plastic flow and failure initiation. The mismatch between matrix-second phase particles consisted of areas like grain boundaries and intergranular and intragranular precipitates induced by quenching, arranged in decreasing order of size. The microstructural heterogeneity and the plastic properties associated with different zones of the materials lead to a complex competition between various failure modes, such as intermetallic decohesion and intergranular and transgranular fracture [3,7].

Generally, an Al-Zn-Mg-Cu alloy can be strengthened by heat treatment or the addition of trace transition elements such as Zr, Ti, Sc, Er, to refine the ingot structure, which can significantly impact the strength, fracture toughness, and processability [8-12]. The primary strengthening mechanisms of conventional Al-Zn-Mg-Cu alloys are precipitation strengthening and solid solution strengthening by Zn and Mg. The precipitation sequence is typically a supersaturated solid solution (SSS) transforming to GP zones then η' phase and η'' phase. This sequence is intricately linked to the percentage of Zn, Cu, and Mg incorporated into the alloy [13]. However, introducing transition elements (e.g., Zr, Ti, Sc) can lead to the formation of different particles (e.g., Al_3(Sc, Zr, Ti)), resulting in grain refinement and dispersion strengthening [14-17]. Hence, optimizing the composition of microalloying elements plays a vital role in controlling the microstructure, which benefits the mechanical properties. Recent studies [14-16] have indicated that Zr and Ti could be a more cost-effective alternative to Sc. Additionally, the solid solubility and diffusion coefficient of Zr and Ti in the Al matrix are lower than that of Sc, which increases the precipitates driving force and volume fraction [14]. The combined addition of Sc and Zr leads to the formation of core-shell $L1_2$-Al_3(Sc, Zr) precipitates, and the coarsening of these particles can be inhibited by the Zr-rich shell [16,17]. With addition of Ti, Al_3(Zr, Ti) particles act as effective nucleation sites during solidification, which may further improve the mechanical properties of the Al-Zn-Mg-Cu alloy [18].

The degradation of alloys is a complex and multifaceted phenomenon involving different failure stages, starting with an incubation phase and advancing through nucleation at damage sites related to the microstructure. These sites undergo gradual changes, eventually resulting in macroscopically observable failure modes. In commercial aluminium alloys, the fracture process is primarily initiated by void nucleation, decohesion, or the fracture of second-phase particles, as extensively documented in various literature [19-22]. The volume fraction, distribution, and shape of second-phase particles and intermetallics can significantly impact the resistance to fracture and damage. The fracture surfaces of Al-Zn-Mg-Cu alloys typically exhibit ductile voids that initiate at large second-phase particles [20]. These large second phase particles are predominantly complex phases containing Cu and Mg, such as orthorhombic S-Al_2CuMg and Mg and Zn rich hexagonal, η-$MgZn_2$, which increase with an increase in alloying addition. Reddy and Rajan [20] found that depending on the deformation characteristics of the matrix, voids, which subsequently grow, are initiated primarily at the interface of the larger inclusions at very low strain (~ 3%), but the larger particles tend to crack after approximately 7% plastic strain in BS L65 Al alloys. Subsequently, the remaining fracture path combines ductile intergranular and transgranular fracture. As a result, it is crucial to understand the role of second-phase particles and intermetallics in the fracture mechanism of commercial Al alloys. Limitations associated with ex-situ tests for this class of materials include the inability to capture real-time deformation and damage processes, which can lead to a lack of understanding of the dynamic mechanisms at play. Therefore, a significant amount of work has been done to directly observe various damage processes in aluminum alloys via in situ observation using a scanning electron microscope (SEM).Some key studies include investigating the effect of grain boundary phases and precipitates on crack growth, the effects of particle geometry through in situ observation and modelling, and examining crack initiation, propagation, and blunting in different series of aluminium alloys [23-31]. Several studies [25-28] have also particularly investigated the effect of Si particles on the damage evolution and fracture behaviour of cast Al-Si-Mg alloys. M. Besterci et al. [30] also investigated the effect of a different volume fraction of Al_4C_3 carbides (8–12 vol%) and Al_2O_3 oxide particles on the fracture evolution. The final rupture mainly involved the linking of the side cracks in perpendicular to the loading direction, occurring in rows of varying density, influenced by the higher volume fractions (12 vol%) of carbide (Al_4C_3) and oxide (Al_2O_3) particles. Recent investigations [7-11] have mostly focused on ex-situ applied stresses to obtain consistent crack growth, with only a limited number of tests conducted within an SEM to underst and fracture behaviour at a smaller scale. However, limited attention has been paid to the in-situ study of the failure mechanisms in multiphase Al-Zn-Cu-Mg alloys at the micrometre scale. In addition, the impact of Al_3M (Zr, Ti and Sc) particles and eutectic phases on the initiation, propagation, and fracture behaviour in the 7xx.xseries alloy has not been studied in detail.

In this study, in-situ observation of tensile testing in a SEM has been used to investigate the failure behaviour of a 7xx.x equivalent alloy, Al-6Zn-2.1Mg-2Cu, with varying fractions of microalloying elements, including Ti, Zr, and Sc. The study aims to discern the relative contributions of matrix, eutectic, and Al_3M precipitates and dispersoids to the material's fracture characteristics. The alloy's microstructure was carefully studied using FESEM, EDS, and XRD to identify the phases present. In-situ FESEM tensile fracture studies were conducted on single-notched specimens to monitor crack propagation, with real-time video footage recorded during testing to identify microscale failure. A microtest controller tracked the 2D strain distribution throughout the experiments. Furthermore, finite element analysis (FEA) was employed to predict stress generation and deformation behaviour at the notch tip. The acquired findings provide valuable insights into the fracture behaviour of Al-Zn-Mg-Cu alloys with varying concentrations of microalloying elements (Ti, Zr, and Sc). These results help to understand how different micro-alloying functions within the matrix improve the mechanical properties. Incorporating in-situ SEM testing provides a direct and detailed examination of the alloy's response to varying microalloying elements, offering crucial information towards optimising future alloy design for enhanced performance in critical applications such as in the aerospace and automotive industries.

4.2 Materials and methods

4.2.1 Materials fabrication

The actual compositions (determined by optical emission spectroscopy, OES, Metavision 1008i3, Metal Power, India) of the experimental alloys are shown in Table 4.1. The composition of alloy A was prepared using commercially available 99.99% pure aluminium, Al-30Cu followed by additions of Zn-5Al, and 99.9% pure Mg. The alloys B and C, and D were produced using further addition of Al-10Ti, Al-10Zrand Al-2Sc master alloys.

The master alloys were melted in a resistance melting furnace using a 3kg graphite crucible preheated at 550°C to prevent moisture absorption. To ensure complete distribution of the Al_3M particles in the alloy, the furnace temperature was held at 745°C ± 5°C for approximately 40 minutes. The melt was continuously stirred for 10 minutes to achieve compositional homogeneity. Prior to pouring, impurities and dissolved hydrogen were eliminated by bubbling pure argon gas at a constant flow rate of 10 l/min for 8–10 minutes, which minimised micro-porosity in the cast blocks. After melting, the slag was removed, and the temperature was adjusted to 700°C. Finally, the melt was poured into a rectangular copper mould with a dimension of 148 mm × 60 mm × 46 mm. The cast blocks were T6 heat-treated. Solutionizing was at 475°C for two hours using a muffle furnace (heating rate of 8°C /min), followed by water quenching. After solutionizing, the four alloys were artificially aged at 120°C for 24 hours.

Table 4.1 Chemical compositions of the cast Al-Zn-Mg-Cu-X alloys (wt. %).

Alloy	Nomenclature	Zn	Mg	Cu	Zr	Ti	Sc	Al
A	Al-6Zn-2.1Mg-2Cu	6.17	2.13	2.01	-	-	-	Bal
B	Al-6Zn-2.1Mg-2Cu-0.4Zr-0.2Ti	6.04	2.11	2.06	0.39	0.22	-	Bal
C	Al-6Zn-2.1Mg-2Cu-0.6Zr-0.4Ti	5.98	2.07	1.96	0.61	0.38	-	Bal
D	Al-6Zn-2.1Mg-2Cu-0.4Zr-0.2Ti-0.25Sc	6.12	2.19	2.03	0.41	0.19	0.25	Bal

4.2.2 Microstructure characterization

Metallographic samples measuring 10 mm x 10 mm x 5 mm were taken from as-cast and heat-treated blocks. They were then mechanically ground with silicon carbide emery papers with grit sizes ranging from 400 to 2500 and then polished with a 0.1 μm Al_2O_3 suspension. Metallographic specimens were etched using Keller's reagent, a solution comprising 2.5% HNO_3, 1.5% HCl, and 1% HF in water for ~25 seconds. An optical microscope (Axio-Imager. M3m, ZEISS, Oberkochen, Germany) and an SEM (Sigma HD, Carl Zeiss, Germany) examined the microstructural morphology of four distinct alloys. The eutectic and Al_3M phase compositions were determined using energy dispersive spectroscopy (EDS) from Oxford Instruments, Abingdon, UK. X-ray diffractometry (XRD) was conducted on a Rigaku Miniflex 600 X-ray diffractometer with Cu Kα radiation (2θ range 10-90°, step size 0.1°, scan speed 10°/min, λ = 1.54 Å) to reconfirm the phases in the matrix. The phase fraction of distinct eutectic and intermetallic phases of different alloys is determined from the peak intensity, which is given by the following expression (eq. 1).

$$Phase\ fraction\ (\%) = \frac{Intensity\ of\ particular\ phase}{Total\ Intensity} \times 100 \quad (1)$$

Transmission electron microscopy (TEM) was conducted using a JEOL JEM F 200 high-resolution transmission electron microscope (HRTEM). The TEM specimens were prepared through twin-jet electro-polishing, using a chemical solution of 30% nitric acid and 70% methanol at -20°C, with an operating voltage of 12V.Hardness evaluation adhered to ASTM E10 standards (Bluestar B-3000 TS) employing a 10 mm ball under a 1000 kgf load, along with phase-specific Vickers hardness testing (Falcon 500, Innova Test) using a 50gf force and a 15-second dwell time per sample.

4.2.3 In-situ tensile testing

In-situ tensile testing was conducted using a Deben Microtest 2-kN testing stage (MTEST2000, GATAN, UK) positioned within the vacuum chamber of FE-SEM (Sigma HD, Carl Zeiss, Germany) [32]. The FE-SEM is equipped with a loading system for applying uniaxial tensile loads. The specimens were horizontally mounted, securely clamped between two jaws with a small precise ball screw and supported on stainless steel slide bearings. Symmetrical movement of the jaws in opposite directions was achieved through a dual-threaded lead screw, ensuring the sample remained centred in the field of view. The micro-test controller controlled a standard speed range of 0.033–0.4 mm/min. An optical encoder-equipped motor with an error of about 1% was used. The sample displacement speed was set at 0.033 mm/min during tensile testing. Figure 4.1a illustrates the geometry of a tensile specimen with 13 ± 0.2 mm gauge length, 5 mm width, and 1 mm thickness. A "U" notch was intricately created in the middle of the gauge length using a wire electrical discharge machine (WEDM) with half of the width to facilitate precise observation of the initiation and propagation of the crack. Due to the stress concentration at the notch during tension, micro-cracks are likely to form near the notch. The graphical user interface (GUI) affords the user comprehensive control over every aspect of the tensile testing machine. SEM images were acquired at loading intervals. Tensile force measurements were obtained through the load cell. Since the load cell (equipped with a microtester) has a maximum tensile force capacity of 2kN, the specimen's gauge dimension, such as minimum gauge width and gauge thickness, was calculated by considering each alloy's tensile strength, which is reported in our previous work [15]. The different Al-Zn-Mg-Cu-X alloys were estimated to break under the machine's maximum tensile force requirement of 2 kN, with a cross-sectional area of 2.5 mm^2 (max. gauge width =2.5mm and gauge thickness=1mm). Furthermore, the macroscopic strain and stress distribution of the designed sample were confirmed by FEA modelling. After in-situ tensile testing, fracture surfaces and crack propagation from beneath the fracture surface were also observed by FESEM.

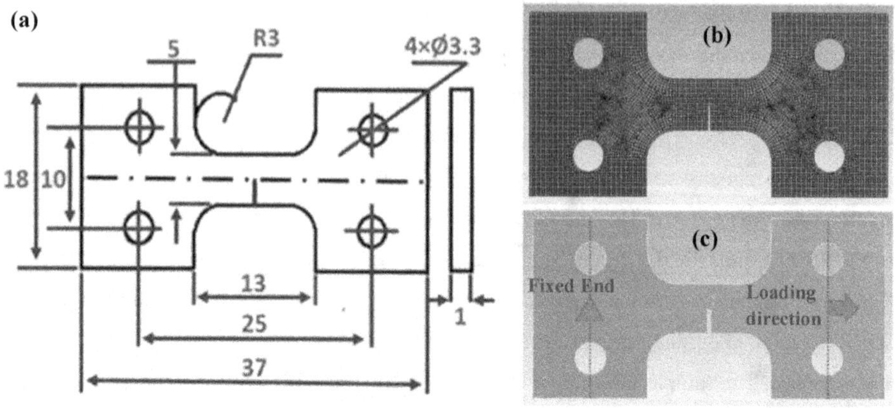

Figure 4.1 (a) Geometry of the miniature tensile specimen (all dimensions in mm) (b, c) meshed specimen and loading direction of the tensile specimen using FEM.

4.2.4 FEM modelling

FEM was also performed to study the typical deformation behaviour and stress distribution on the gauge length. ANSYS^{TM} Workbench (version 19.2) simulates tensile behaviour with an explicit dynamics module [33]. A 3D CAD model of the tensile specimen was drawn using SolidWorks software. The investigation centred on the precisely determined dimensions and geometry of the small tensile specimen used for in-situ tensile testing, as depicted in Figure 4.1. The solid CAD model was imported into the ANSYS workbench, and the required properties, such as material properties, applied loading, and constraint information, were assigned. In this study, the material used was cast 7xx.x series Al-Mg-Cu-Zn alloy. The properties of the material are given in Table 4.2. Fine meshing was done with 0.25-mm fine-sizing meshing for the specimen. The model consisted of 40515 nodes and 31128 linear elements, where mostly 4-node quadrilateral elements were used to complete the mesh (Figure 4.1b). In the next stage, the load and boundary conditions were applied as detailed in appendix B. One end of the specimen was kept fixed, and the other end was assigned to move with a constant speed of 0.033 mm/min (Figure 4.1c). The tests were conducted using the same parameters as used in in-situ tensile testing.

Table 4.2 Material properties of cast 7xx.x series alloys used for simulation.

7xx.x T6	
Physical properties	
Density (ρ)	2.81 g/cm^3
Mechanical properties	
Young's modulus (E)	71.7 GPa
Tensile strength (σ_t)	574 MPa
Elongation (ε) at break	5.90%
Poisson's ratio (ν)	0.33
Hardness—Brinell	179 BHN
Thermal properties	
Melting temperature (Tm)	750 K
Thermal conductivity (*k*)	130–150 W/m*K
Linear thermal expansion coefficient (α)	$2.36*10^{-5}$ K^{-1}
Specific heat capacity (c)	714.8 J/kg*K

4.3 Results

4.3.1 Material characterization

Figure 4.2 exhibits the typical optical microstructures of all as-cast alloys with and without adding different microalloying elements such as Ti, Zr, and Sc. The microstructural morphology showcases the characteristic as-cast eutectic structure with continuous network morphology along the grain boundaries. Figure 4.2a (high magnification micrograph) identifies two prominent characteristics of eutectic phases. The first feature is represented by a clear black-scale contrast between η-$MgZn_2$ and S-Al_2CuMg phases at the grain boundaries, highlighted by the red and yellow arrows. The second feature is represented by dot-shaped T-$Al_2Zn_3Mg_3$ phases, which have a grey-scale contrast and are shown by the blue arrow, as observed in previous studies [15]. Upon the addition of 0.2 wt.% Ti and 0.4 wt.% Zr, the α-Al cell size decreases (Figure 4.2b) from 62 ± 1.4 μm to 48 ± 1.3 μm. Small and thin rod-like Al_3(Zr, Ti) particles in alloy B with an average length ranging from 5 μm to 30 μm. However, the addition of Ti (0.4 wt.%) and Zr (0.6 wt.%) leads to the formation of large bulky Al_3Zr and Al_3(Zr, Ti) intermetallics, with an average length of 120 μm. By comparing alloy C and alloy B, it appears that the large intermetallics cannot restrict the coarsening of the α-Al cell size (51 ±1.6 μm) (Figure 4.2c). Therefore, alloy B was chosen to further add 0.25 wt.% Sc, designated as alloy D, and the grain structure transformed from dendritic to equiaxed with an average α-Al cell size of 34 ± 1.1 μm, (Figure 4.2d). The microstructural morphology of the eutectic S-Al_2CuMg, η-$MgZn_2$ and T-$Al_2Zn_3Mg_3$ in the four as-cast alloys exhibit no notable differences.

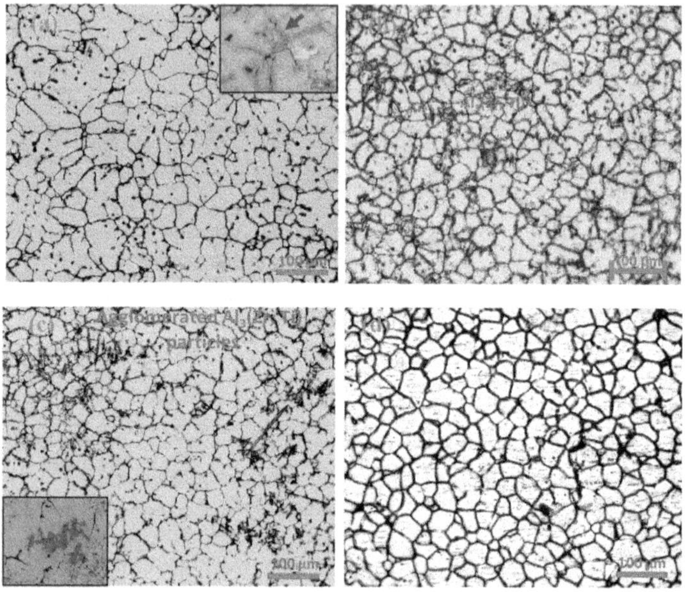

Figure 4.2 Optical microstructures of as-cast (a) alloy A shows η-$MgZn_2$, S-Al_2CuMg and T-AlZnMgCu phases indicated by red, yellow and blue arrows (b) alloy B shows small size Al_3(Zr, Ti) dispersoids indicated by yellow arrow (c) alloy C shows large agglomerated Al_3(Zr, Ti) dispersoids (d) alloy D.

After that, the alloys are subjected to a T6 heat treatment cycle, as described in section 2.1. Figure 4.3(a-d) displays SEM micrographs of T6 heat-treated specimens. In the microstructure of alloy A, most of the eutectic phases are dissolved after solution treatment, but a small volume fraction (5%) of the lamellar S-Al_2CuMg phase and η′-$MgZn_2$ are distributed along the grain boundaries. As shown in Figure 4.3c, irregular bulk Al_3(Zr, Ti) intermetallics are located in the middle of the matrix phase. This study also observed ultra-fine Al_3(Sc, Zr) precipitates after T6 treatment (Figure 4.3d). Figure 4.4a and 4.4b are bright field TEM micrographs depicting the variation in precipitate distribution along the grain boundary in T6-heat treated alloy D. In alloy D (Figure 4.4a), continuous and high-density rod- and plate-like metastable η' precipitates with grain boundary precipitate (GBP) morphology are observed. These precipitates, nucleate uniformly, are metastable η' phases with a composition close to $MgZn_2$ as observed in Figure 4.4a and 4.4b with corresponding selected area diffraction patterns having zone axis [100].To identify the large agglomerated dispersoids, SEM-EDS maps (Figure 4.4c) show distinct concentrations of Ti, Zr, and Sc in the Al_3M phase, while Cu and Mg are concentrated in the grain boundary region in the heat-treated alloy D specimen. Furthermore, X-ray diffraction analysis is performed to reconfirm the formation of eutectic and dispersoids in both as-cast and T6 heat-treated specimens.

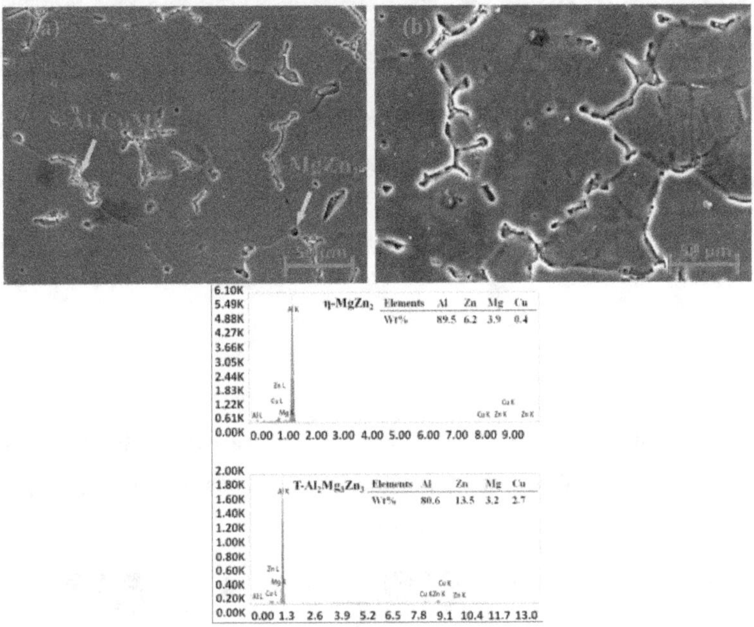

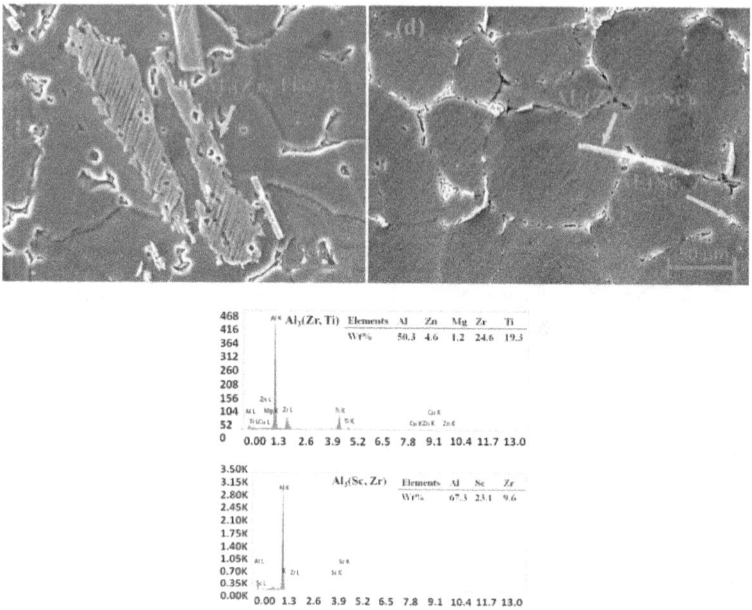

Figure 4.3 Secondary electron micrographsof T6 heat-treated (a) alloy A (b) alloy B (c) alloy C and (d) alloy D with corresponding identified phases by energy dispersive spectroscopy (EDS).

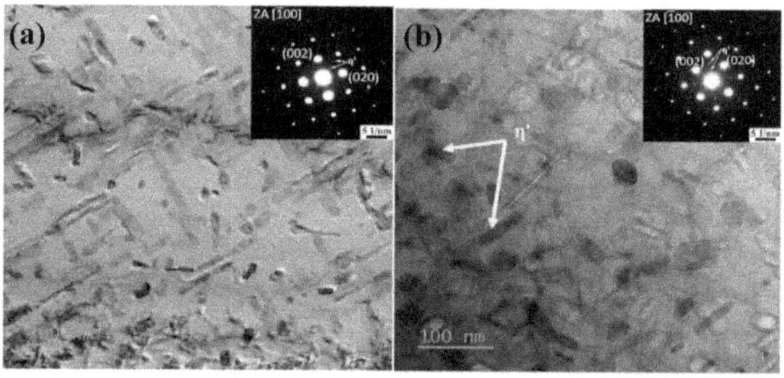

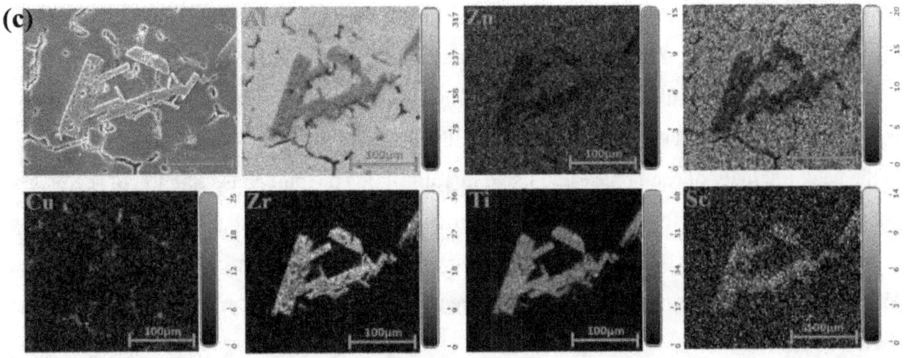

Figure 4.4 TEM micrographs of(a) High density η' precipitate distribution with selected area electron diffraction (SAED) pattern and (b)high magnification TEM micrograph showing η′phases with corresponding SAED pattern. (c) SEM-EDS area mapping of alloy D after T6 heat treatment.

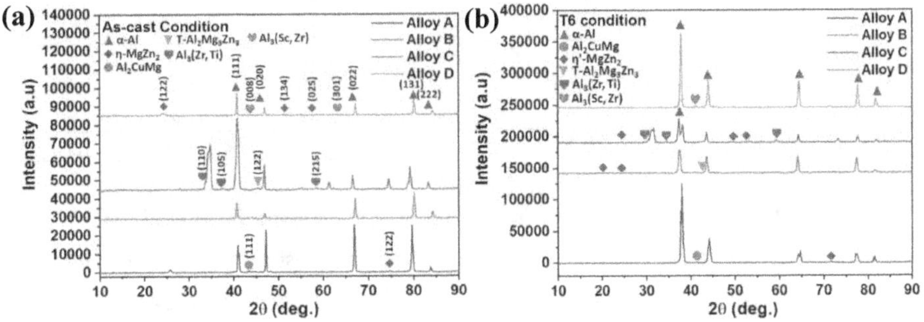

Figure 4.5 XRD analyses alloy A, B, C and D in (a) as-cast (b) T6 heat treated condition.

The corresponding XRD patterns are displayed in Figure 4.5. Primary diffraction peaks are mainly attributed to the α-Al, η and metastable η′ phases. The primary dendrites of α-Al in the alloy exhibit Bragg diffraction peaks along the (111), (020), (022), (131), and (222) crystallographic planes. However, it is noted that the standard diffraction peaks of the η-$MgZn_2$ phase exhibit slight deviations for different alloy systems with preferred (122), (134), and (025) crystallographic planes. A small peak corresponding to the Al_2CuMg phase is evident in four different as-cast and heat-treated alloys oriented along $(011)_S$ // $(111)_{\alpha\text{-Al}}$ with α-Al. Alloy B did not show peaks corresponding to Al_3(Sc, Zr), and Al_3(Zr, Ti) phases. However, a tiny peak is observed in alloy C, likely due to a higher fraction of microalloying elements than in alloy B. After T6 heat treatment, the XRD peak corresponding to the η-$MgZn_2$ phase decreases and almost disappears in alloys A, B, C and D. The peak intensity of secondary S-Al_2CuMg phase varies, as evidenced by the microstructures resulting from heat treatment. The XRD peaks of $Al_3(Zr, Ti)$, and $Al_3(Sc, Zr)$ phases remain unchanged throughout the heat treatment process. Furthermore, the XRD diffraction pattern of alloy D confirms the existence of $Al_3(Sc, Zr)$. The phase fraction of distinct eutectic and intermetallic phases of different alloys is calculated based on the peak intensity of each phase and listed in Table 4.3. The total phase fraction of secondary eutectic phases varies from 6.5% to 1.5% in four different as-cast and heat-treated alloy systems, confirming the dissolution of secondary phases. The phase fraction of $Al_3(Zr, Ti)$ and $Al_3(Sc, Zr)$ dispersoids and precipitates increases with increasing microalloying contents (Table 4.3).

Table 4.3 Phase fraction (%) of as-cast and T6 heat-treated alloys.

Condition	Alloy	α- Al	η/ η'- $MgZn_2$	S- Al_2CuMg	T- $Al_2Mg_3Zn_3$	$Al_3(Zr, Ti)$	$Al_3(Sc, Zr)$
As-Cast	A	93.59	3.86	1.52	1.03	-	-
	B	93.13	3.73	1.49	1.19	0.46	-
	C	92.93	3.96	1.42	0.86	0.83	-
	D	92.57	3.71	1.54	1.16	0.49	0.53
T6 heat-treated	A	98.35	0.51	0.68	0.46	-	-
	B	98.08	0.56	0.61	0.33	0.42	-
	C	97.51	0.63	0.67	0.38	0.81	
	D	97.56	0.59	0.52	0.32	0.46	0.55

4.3.2 Deformation behaviour of a notched tensile specimen using FEM analysis

Figure 4.6 represents the distribution of equivalent elastic and plastic strain and equivalent (von Misses) stress at 0.69 mm displacement (Figure 4.6a) and after fracture (Figure 4.6b) calculated using the explicit dynamics model in ANSYS workbench with cast 7xx.x series Al-Zn-Mg-Cu material data (as available in the materials database in ANSYS), with a composition close to the alloys selected for this experimental study. This simulation only represents macroscopic tensile properties, which validate stress and strain distributions on tensile specimen geometry with a notch considered for study. It can be assumed that the macroscopic stress and strain on the sample may be applicable for every composition. The results indicate that the elastic strain (Figure 4.6a1), and plastic strain (Figure 4.6a2) are maximum at the tip of the notch with uniform distribution along the gauge width except for higher strain in a V-shape away from the notch. The equivalent (von Mises) stress (Figure 4.6a3) is distributed equally in the necked zone of the tensile specimen. Since the strain is maximum near the edges of the notch, the fracture crack initiated from the edge of the notch (the grain boundary and precipitates may act as crack nucleation points). The uniformity in the strain distribution in the two arms of the V-shape strain distribution and the uniform distribution of equivalent stress indicates that the tensile sample designed and fabricated for the study is perfect for conducting experiments to understand in-situ tensile deformation behaviour. The sample condition after fracture (Figure 4.6b) indicates that the crack propagates along the path of high strain concentration. A similar fracture morphology has been observed experimentally macroscopically (Figure 4.7-4.10). Further microscopic fracture analysis considering precipitates and grain boundaries was conducted by in-situ fracture observation using electron microscopy.

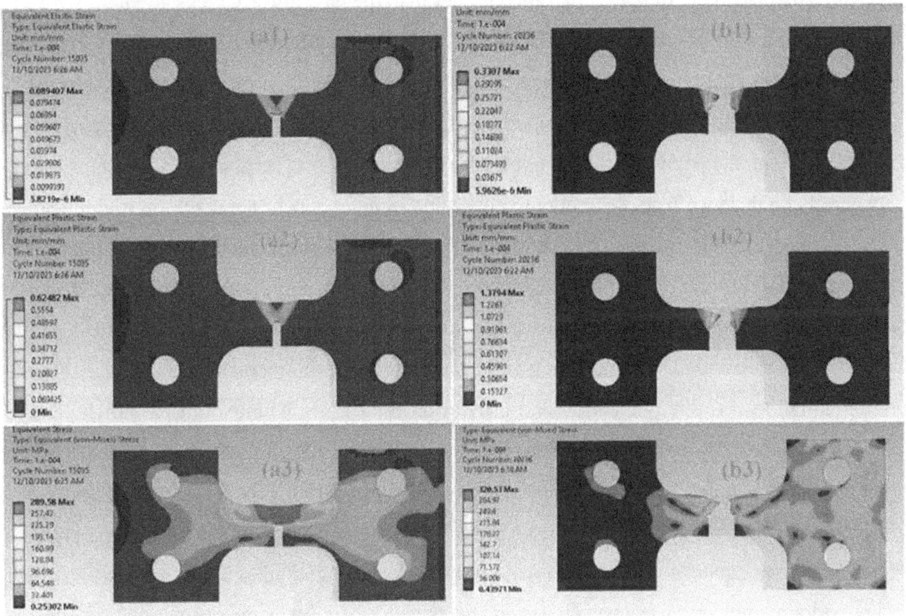

Figure 4.6 (a1-a3) shows Elastic strain, plastic strain, and von Mises stress distribution at 0.69 mm displacement and (b1-b3) shows plastic strain, and von Mises stress distribution after fracture.

4.3.3 In-situ observation of deformation behaviour of Al-Zn-Mg-Cu-X alloys

In-situ uniaxial tensile deformation was carried out on the notched specimens (Figure 4.1) to investigate the mechanical response of different alloys. Figure 4.7 represents the in-situ deformation characteristics and fracture behaviour of the alloy A specimen. The failure in the specimen occurred at a peak load of 432 N without significant necking, as shown in Figure 4.7a. The main crack, initiated at the notch, encounters grain boundary intermetallic phases (Figure 4.7b and 4.7c) during the propagation stage (stage II) through the grain boundaries (Figure 4.7d). Re-nucleation of cracks at the shrinkage pores is also observed. The fracture path displays a macroscopically straight trajectory across the gauge section even while exhibiting interaction with microstructural details (Figure 4.7e). In some parts of the matrix, the crack path's meandering features are related to crack tip deflection around the eutectic colonies of the notched tensile specimen. The in-situ tensile examinations, therefore, reveal the distinctive behaviour exhibited by various brittle components during fracture.

The load vs. displacement plot of alloy B, illustrated in Figure 4.8a, for the alloy with the addition of 0.2 wt. %Ti and 0.4 wt. % Zr experienced failure at a nominal average load of 965 N, with a displacement of 0.42 mm, much higher than the base alloy. Cracks are mainly formed by the decohesion of the eutectic phases at the grain boundaries (Figure 4.8b). Comparing Figure 4.7 and 4.8 show that the crack initiation and propagation mechanisms are consistent for alloys A and B. During the initial loading phase (Stage-I), minimal evidence of deformation is observed in the vicinity of the notched specimen. Subsequently, multiple microcracks appeared at a distance from several eutectic grain boundaries near the crack tip region (Stage-II, Figure 4.8c). The microcracks tend to originate from the eutectic dendrite boundaries (Figure 4.8d) and were randomly distributed up to a distance of ~150 μm from the notch. These microcracks predominantly align along the lines of maximum shear stress. With a further increase in applied load, the nucleation and accumulation of microcracks at a distance from the notch coincide with the nucleation and growth of cracks originating from the notch surface. The convergence of microcracks determines the path of failure and leads to complete fracture (Figure 4.8e). In alloy B, smaller clusters of Zr and Ti-rich Al_3(Zr, Ti) dispersoids also influence the crack formation. Smaller Al_3(Zr, Ti) particles (approximately 5 μm and having a fine rod-like or square shape) cause localised crack arrest and tend to blunt the growing crack. This phenomenon leads to an increase in the strength of alloy B in comparison to alloy A. However, in the early stages of damage development in Al-6Zn-2.1Mg-2Cu-X cast alloys, microcracks start mainly by breaking up Cu and Mg-rich particles.

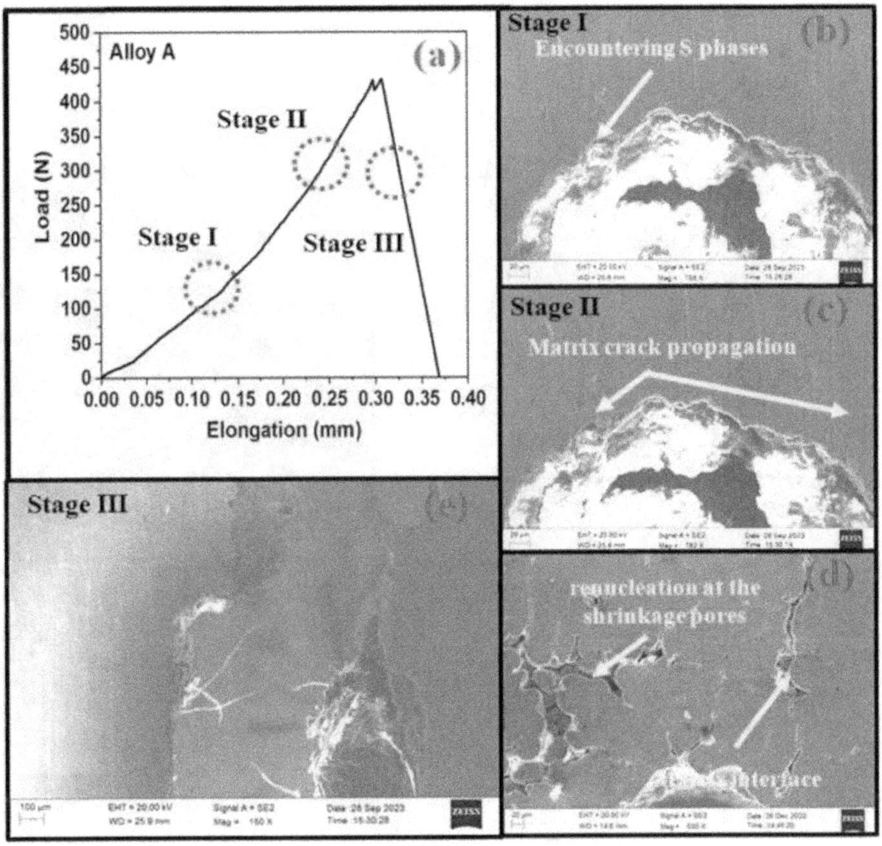

Figure 4.7 In-situ observation of alloy A in SEM under load (a) load vs. elongation plot of the U notch specimen with corresponding fracture steps marked with stage I, II and III (b) crack initiation at the tip of the notch (c) crack propagation through the eutectic phases (d) re-nucleation at the shrinkage pores in the matrix (e) final fracture of alloy A.

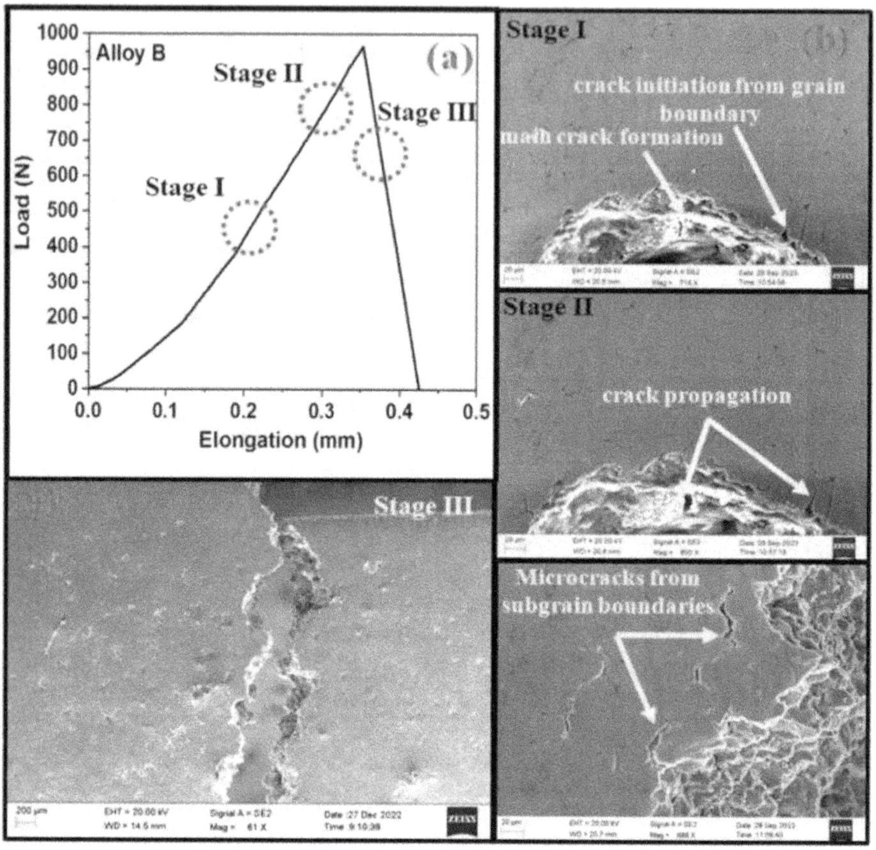

Figure 4.8 In-situ observation of alloy B in SEM under load (a) load vs elongation plot of the U notch specimen with corresponding fracture steps marked with stage I, II and III (b) crack initiation at the tip of the notch (c) formation of main crack and microcracks from different grain boundary eutectic phases in the matrix (d) enlarged view of formation of different micro-cracks and their coalescence (e) final fracture of alloy B.

Further addition of microalloying elements to 0.4 wt.% Ti and 0.6 wt.% Zr, in alloy C, leads to large-size dispersoids ~ 120 μm in length (Figure 4.3c). The load vs. elongation diagram of alloy C shows different failure stages (Figure 4.9a). Figure 4.9(b-e) shows how cracks move through alloy C. Interestingly, the crack tends to initiate at the large $Al_3(Zr, Ti)$ phases directly without any noticeable deformation in the matrix, thereby not efficiently consuming energy during crack propagation. Owing to the limited deformation of the matrix, the stresses are likely to be concentrated near the interface of the matrix and the large agglomerated $Al_3(Zr, Ti)$ phase. The failure mainly occurred in the middle section of the intermetallic layer (Figure 4.9d). The load-displacement curve also exhibits a load drop, signifying a sudden jump in crack propagation due to the fracture of the large intermetallic phase (Stage-I). However, after this, the matrix strengthened further due to crack bridging and crack meandering effects, as seen in alloy C (Stage-II in Figure 4.9c).

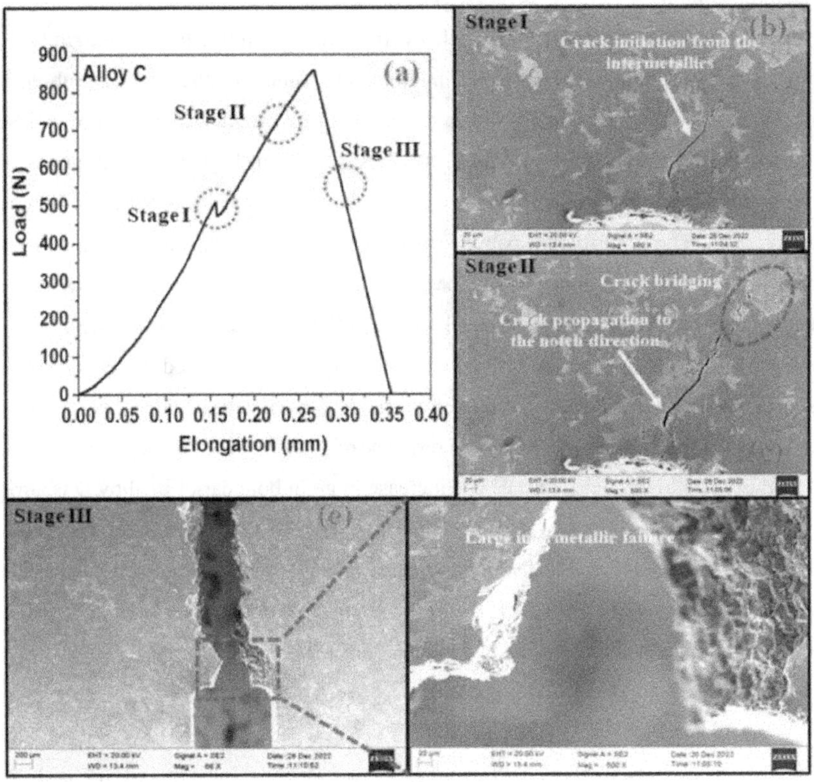

Figure 4.9 In-situ observation of alloy C in SEM under load (a) load vs elongation plot of the U notch specimen with corresponding fracture steps marked with stage I, II and III (b) crack initiation from the Zr, Ti rich dispersoids(c) crack propagation, crack bridging and crack meandering in different large agglomerated dispersoids(d) failure and corresponding fracture surface (e) final fracture of alloy C.

The higher phase fraction (Table 4.3) of grain boundary and $Al_3(Ti, Zr)$ dispersoids [18]in alloy C leads to a rapid fracture process due to the material's low plasticity. When the dispersoids are approximately 10–25 μm in specific specimen locations, cracks encounter the interface between the $Al_3(Zr, Ti)$ phases and α-Al matrix. Then the cracks shift the direction before extending along the grain boundaries of the S-Al_2CuMg lamella. This crack meandering effect consumes energy during crack propagation, leading to further deformation in the matrix. Consequently, attempting to propagate the crack along the lamella becomes a formidable task. Notably, the S-Al_2CuMg lamella is more brittle than the fine Al_3Zr or $Al_3(Zr, Ti)$ dispersoids. As a result, a new crack emerges and extends across the grain boundary lamella rather than propagating along it.

After reaching a specific load of 858 N, alloy C fractures, exhibiting lower strength than alloy B, due to the formation of the distinct agglomerated intermetallic phases. Subsequently, 0.25wt.% Sc was introduced to alloy B in this study to investigate the impact of $Al_3(Sc, Zr)$ precipitates on the propagation of cracks. It is observed from Figure 4.7, 4.8, and 4.10 that nearly identical crack patterns emerged in alloys A, B, and D despite significant differences in their structures. The deformation patterns of alloy D indicate that crack initiation occurred primarily through decohesion of S-Al_2CuMg, occasionally complemented by the rupture of large grain boundary particles. The crack paths initially aligned with the locations of the microcracks in front of the tip until a second large crack formed in close proximity. Subsequently, the interaction between the stress fields of the cracks dictated their paths. Figure 4.10a shows the typical load vs. elongation response of alloy D, which demonstrates the highest peak load of the alloy systems tested. The increase in grain boundaries in alloy D is attributed to adding Sc and a small fraction of Ti and Zr. The presence of $Al_3(Sc, Zr)$ precipitates, with a size of approximately 1-8 μm, in the microstructure makes it challenging to observe their effect during crack propagation. However, the study found that the very fine 3 μm $Al_3(Sc, Zr)$ dispersoids blunt the crack tip for almost 300 seconds (Figure 4.10d and 4.10e).

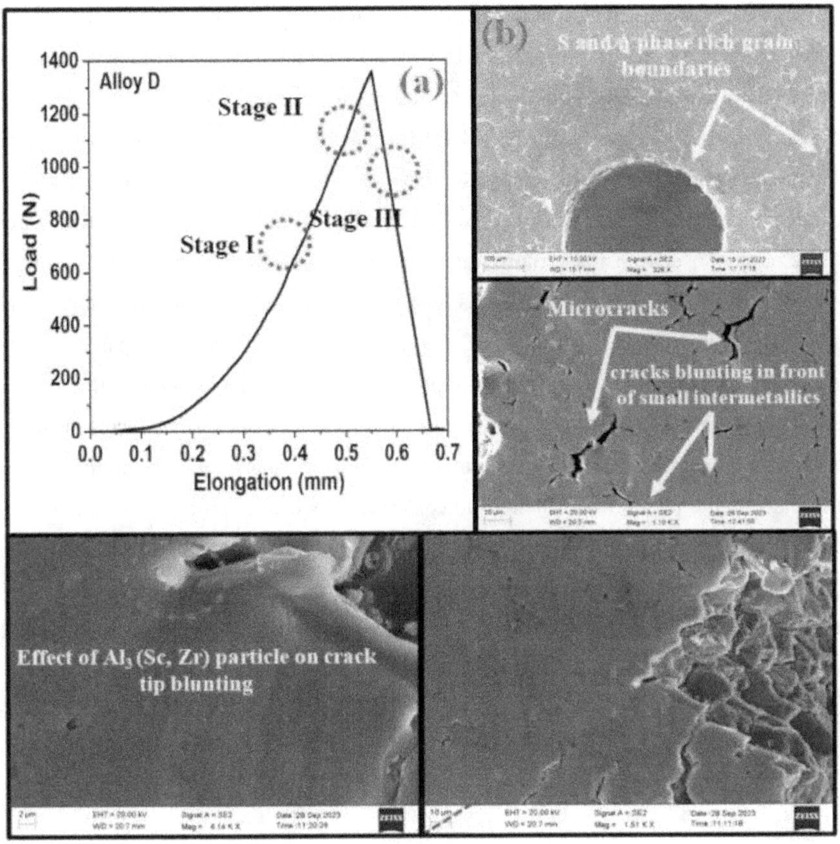

Figure 4.10 In-situ observation of alloy D in SEM under load (a) load vs. elongation plot of the U-notched specimen with corresponding fracture steps marked with stage I, II and III (b) microstructure around the notch tip with fine dispersoids distribution (c) cracks blunting in front of small dispersoids and grain boundary cracks (d) effect of Al_3(Sc, Zr) particle on crack tip blunting (e) enlarged view of Al_3(Sc, Zr) particle.

4.3.4 Fractographic examination

The fracture surfaces of the experimental alloys post-failure are analysed using SEM and are shown in Figure 4.11. The fracture features predominantly confirm quasi-brittle fracture consistent with the load vs. elongation plots of the U-notched specimens. The fracture surfaces of alloys A, B, C and D also reveal areas of porosity that encompass approximately 9% of the fracture surface area. The SEM image of the entire surface of alloy A in Figure 4.11a revealed minimal areas of porosity. The fracture surface of alloy A exhibits characteristics of quasi-brittle fracture, accompanied by a distinct ductile damage zone with dimple features and tear ridges (Figure 4.11a1 and 4.11a2). Alloy B displays similar quasi-brittle fracture regions with larger voids (Figure 4.11b). Both ductile and cleavage facet regions are identified by yellow boxes and enlarged in Figure 4.11b1 and 4.11b2, respectively. A minor section of the specimen exhibits transgranular fracture, likely to increase ductility. Fine and deep dimples with secondary phases are seen within the large voids in alloy B. As the Ti and Zr content increased to 1 wt.%, alloy C demonstrated more brittle fracture, as shown in Figure 4.11c. Figure 4.11c also shows fractured $Al_3(Zr, Ti)$ phases, and grain boundary S-Al_2CuMg lamella. These features on the fracture surface in alloy C suggest its higher brittleness compared to alloys A and B. Noticeably, an increase in finer dimples was observed in alloy D, possibly due to the presence of $Al_3(Sc, Zr)$ dispersoids, in addition to the presence of cleavage facets, which helps it achieve the maximum strength among all selected alloys. Many equiaxed dimples in Figure 4.11d indicate the initiation of micro-void coalescence, which is probably related to the fine dispersoids due to Sc modification. In addition, small holes appear due to the clustering of $Al_3(Zr, Ti)$ dispersoids, while tear ridges are still present on the fracture surfaces of alloy D. Comparing alloys C and D, it is observed that the fracture surfaces of alloy C indicate that the hard inclusion of intermetallic has failed either by fracture or by debonding. Among the examined alloys, finer voids and void sheets are mostly observed in the case of alloy D, whereas shallow and deep dimples can be observed in alloy B (Figure 4.11b and 4.11d).

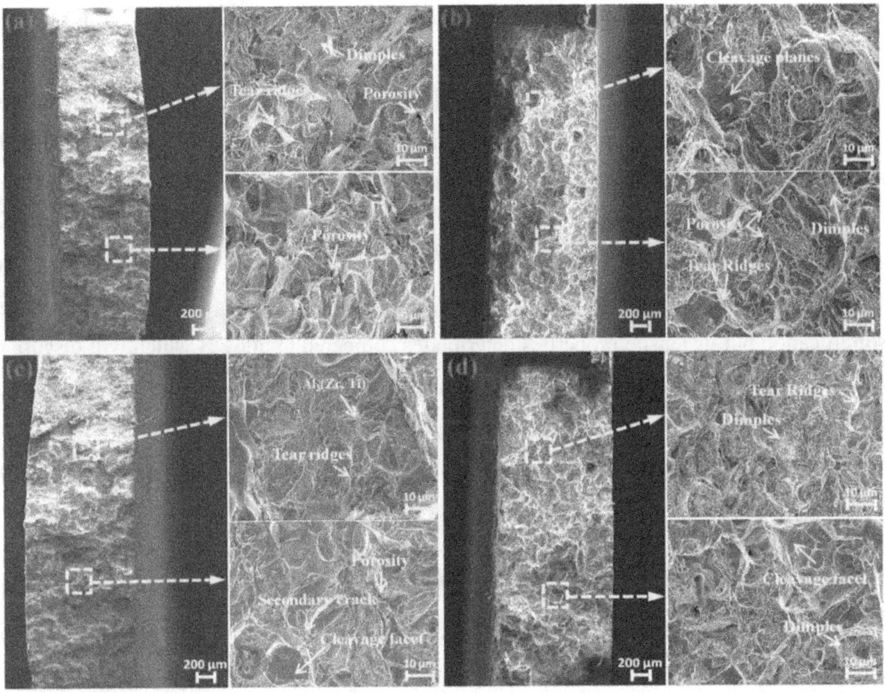

Figure 4.11 Secondary electron micrographs. (a, b, c, d) shows the fracture surface of alloys A, B, C, and D of a single U-notched specimen during in-situ tensile testing (a1, b1, c1, d1), which corresponds to a primarily quasi-brittle fracture zone observed with large Ti and Zr-rich dispersoids, whereas (a2, b2, c2, d2) are the corresponding ductile fracture zones with variation in void morphologies and features.

4.4 Discussion

4.4.1 Microstructures of Al-6Zn-2.1Mg-2Cu-X alloys

The presence of Ti, Zr, and Sc particles restrained the movement of alloying elements into the liquid phase and impeded element separation within the solidification microstructure. This resulted in the accumulation of elements within the α-Al matrix, facilitating their dissolution during the solutionizing [16, 18]. Furthermore, micro-alloys introduced defects that increased vacancy concentration, promoting the diffusion of solute elements (e.g., Zn, Mg, and Cu) during solutionizing. This simultaneous increase in vacancies initiated nucleation, leading to significant precipitation influenced by micro-alloys and T6 heat treatment. In comparison to alloy A, alloy D exhibits a notable reduction in grain size, as shown in Figure 4.2a, 4.2d, and 4.3a, 4.3d. This can be attributed to the positive refining effect of scandium, which forms $L1_2$-$Al_3(Sc, Zr)$ dispersoids [17], as shown in Figure 4.3d. The addition of Ti and Zr has a smaller impact on grain refinement than Sc, as shown in Figure 4.2b, 4.2c, and 4.3b, 4.3c. Nucleation of α-Al does not occur in the substrate of the bulky phases, leading to a lack of significant grain refinement in alloys B and C. The addition of Zr, Ti and Sc also helps to mitigate the tendency for ingot cracking and enhances the mechanical properties of investigated alloys [11]. Notably the hardness of alloy D increases significantly of ~95% compared to base alloy A listed in supplementary Table A1 aligning with the Hall-Petch relationship where grain size reduction and hardness enhancement occur through micro-alloy addition. The hardness of alloy C exhibits ~146 BHN due to the presence of large $Al_3(Zr, Ti)$ intermetallics formed during solidification but its tensile properties deteriorates compared to alloy B which will discuss in next section. It is noteworthy that the concentrations of Zr, Ti, and Sc are relatively lower compared to other key elements such as Zn, Mg, and Cu, even though the overall microalloying concentration varies from 0.6 wt.% to 1 wt.%. However, previous studies have observed the formation mechanism of these $L1_2$ structures, which showcase a distinctive layered composition consisting of a core enriched in $Al_3(Sc, Zr)$, surrounded by layers rich in Sc and Zr, i.e., $Al_3(Sc, Zr)$ + α-Al + $Al_3(Sc, Zr)$. It has been suggested that the primary phases forming during solidification are either Al_3Sc or Al_3Zr, originating from pre-existing particles in the melt [14]. This highlights the behavior of Sc and Zr partitioning during the solidification of primary $Al_3(Sc, Zr)$ particles from the melt, which is governed by the long-range diffusion of Sc compared to Zr within the α-Al matrix. It is also noted that primary alloying elements like Mg, Cu, and Zn contribute to the nucleation of primary $Al_3(Sc, Zr)$ by generating eutectic η′-$Mg(Cu, Zn)_2$ or prototype η′-$MgZn_2$ and S-Al_2CuMg phases at the grain boundaries [16]. The presence of these 20–30 μm lath-like secondary irregular-shaped S-Al_2CuMg eutectic phases is illustrated in Figure 4.3. Interestingly, increasing the fraction of Zr and Ti percentage in the Al-Zn-Mg-Cu (designated as alloy C) shows a poisoning effect of these microalloying elements, as illustrated in Figure 4.2c and 4.3c. The concentration of these microalloying elements at the solid/liquid interface accelerates the premature formation of bulky $Al_3(Zr, Ti)$ intermetallic compounds, depleting these alloying elements and diminishing constitutional supercooling near the interface [7-11]. Due to this tensile strength of alloy C reduces ~50 MPa compared to optimum micro-alloyed (Zr + Ti) alloy B. This is affirmed by Figure 4.3c, which reveals large-sized Ti and Zr-rich intermetallics predominantly formed in the matrix. In addition, the presence of Zr in Al alloy can hinder the refinement effect of Ti due to the interaction between Zr atoms and Al_3Ti particles. This interaction impedes the formation of α-Al on Al_3Ti, resulting in a detrimental effect when Zr and Ti are introduced at a large volume fraction. These particles also induce η phase precipitation and segregation, leading to a decrease in the strength of the Al-Zn-Mg-Cu alloy, as shown in Figure 4.9a. However, according to the Zener pinning mechanism, a substantial volume fraction of small coherent precipitates is essential to counteract the driving force for boundary migration and achieve the highest strength [7]. However, the precipitation strengthening phenomenon is attributed to the presence of small-size coherent $L1_2$-Al_3(Zr, Ti) for alloy B and both coherent $L1_2$-Al_3(Sc, Zr, Ti) and Al_3(Sc, Zr) phases for alloy D.

4.4.2 Micro-mechanisms of fracture behaviour of Al-6Zn-2.1Mg-2Cu-X alloys

The micro-mechanism of fracture observed in the Al-6Zn-2.1Mg-2Cu-X alloy, similar to other high-strength aluminium alloys such as Al-Ce and AA2124/SiC alloys, predominantly reveals a quasi-brittle fracture (Figure 4.11) with limited ductility [24, 36]. The tensile strength of these alloys is mainly determined by the cleavage fracture mechanism and particle/intermetallic decohesion [24, 25]. Additionally, the limited ductility of the matrix is a result of the formation of voids at the interface with second-phase particles and the intermetallics and dispersoids, which undergo nucleation, growth, and coalescence [21, 24, 25, 37].The strength of alloy B is comparatively higher than that of alloy C due to the more optimum addition of microalloying elements. Approximately 0.4% Al_3(Zr, Ti) phases are formed in the matrix of alloy B, which helps to accumulate dislocations near the interface of the dispersoids, increasing the strength [38, 39]. These dislocations serve as nucleation sites for precipitates during 24-hour ageing at 120°C. In a related system, Liu et al. [40] observed a significant consumption of dislocations during ageing, promoting the nucleation of strengthening precipitates. This feature of increased solid solubility and dislocation density in alloys can lead to increased microstrain within grains, facilitating the formation of strengthening precipitates during ageing. The strength in alloys C and D originates from the interaction between the high levels of dislocation density. The limited ductility in alloy C is because the holding temperature during casting is not high enough (i.e.; 745°C) to fully dissolve the 1 wt.% of Ti and Zr in the alloy. This temperature is below the dissolution range required to form a homogenous solution for these alloying elements, particularly Ti and Zr, which are known to form stable intermetallic phases like $D0_{22}$ or $D0_{23}$ Al_3(Zr, Ti). This study identified two forms of the Al_3M phase: the metastable cubic $L1_2$ form solution treatment which actually resists crack propagation as observed in alloy B and D, and the stable $D0_{23}$ form during solidification i.e; detrimental for mechanical properties as observed in alloy C. Karpets et al. have reported that the dissolution limits and substitution mechanisms of Ti and Zr in Al alloys, noting that temperatures approaching or exceeding 800°C are often needed to prevent Al_3(Ti, Zr) formation [41]. The presence of such intermetallics can contribute to localized stress concentrations, as they act as brittle phases within the alloy matrix as observed in alloy C. This effect can impede plastic deformation and thus reduce the ductility of the alloy [42, 43]. The coarse $Al_3(Ti, Zr)$ intermetallics form (approximately 120 μm in size), leading to early fractures (Figure 4.9). It was observed in alloy C that the coarse dispersoids crack before any significant deformation in the matrix (Figure 4.12a), allowing the main crack to move unhindered (Figure 4.12b). Figure 4.12c highlights typical intergranular cracking in alloy A, which has no microalloying additions. The ductility of alloy A may be attributed to the obstruction of cracks at the grain boundaries by the presence of eutectic phases, as depicted in Figure 4.12d. Alloy B shows higher strength and ductility as compared to alloy A because crack growth is significantly retarded by the $Al_3(Ti, Zr)$ particles varying from 5 μm to 25 μm in size, as shown in Figure 4.12e. Alloy D with addition of Sc shows maximum resistance to crack propagation.

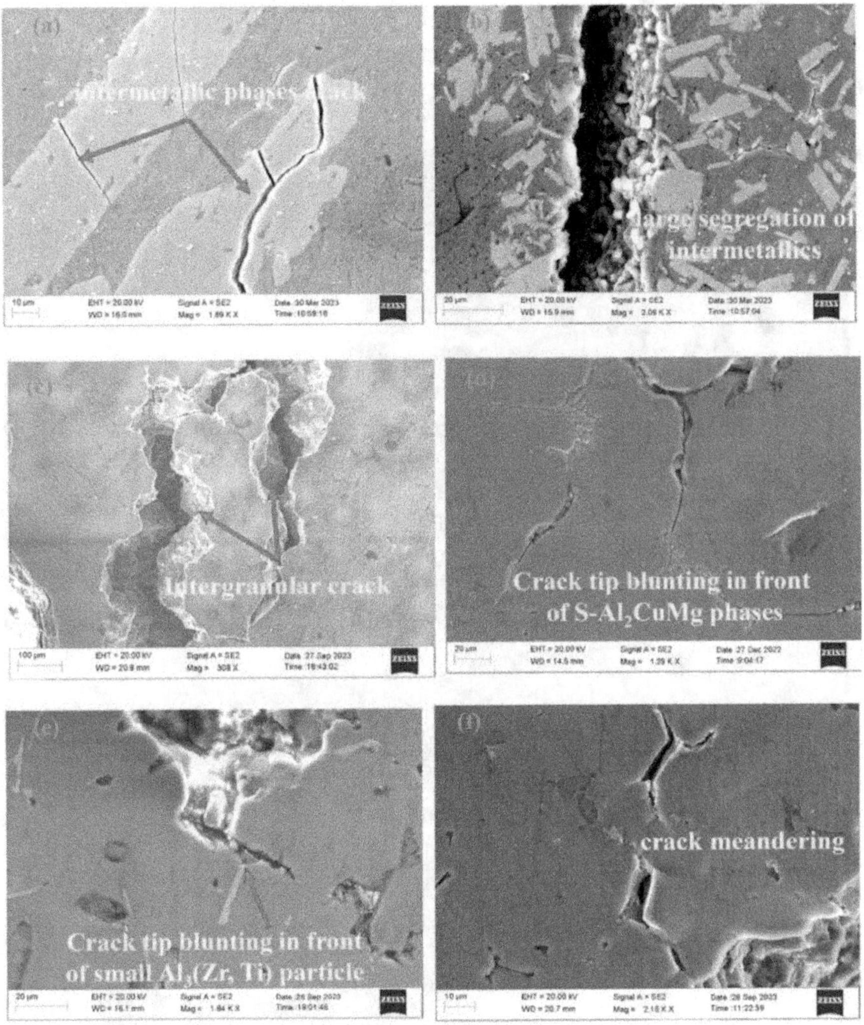

Figure 4.12 Secondary electron micrographs. (a) Al_3(Zr, Ti) phases crack in alloy C (b) fracture due to segregation of large segregation of intermetallics in alloy C (c) intergranular crack in alloy A (d) crack tip blunting in front of grain boundary S-Al_2CuMg phases in alloy A (d) crack tip blunting infront of small Al_3(Zr, Ti) intermetallics in alloy B (d) crack meandering phenomena in alloy D.

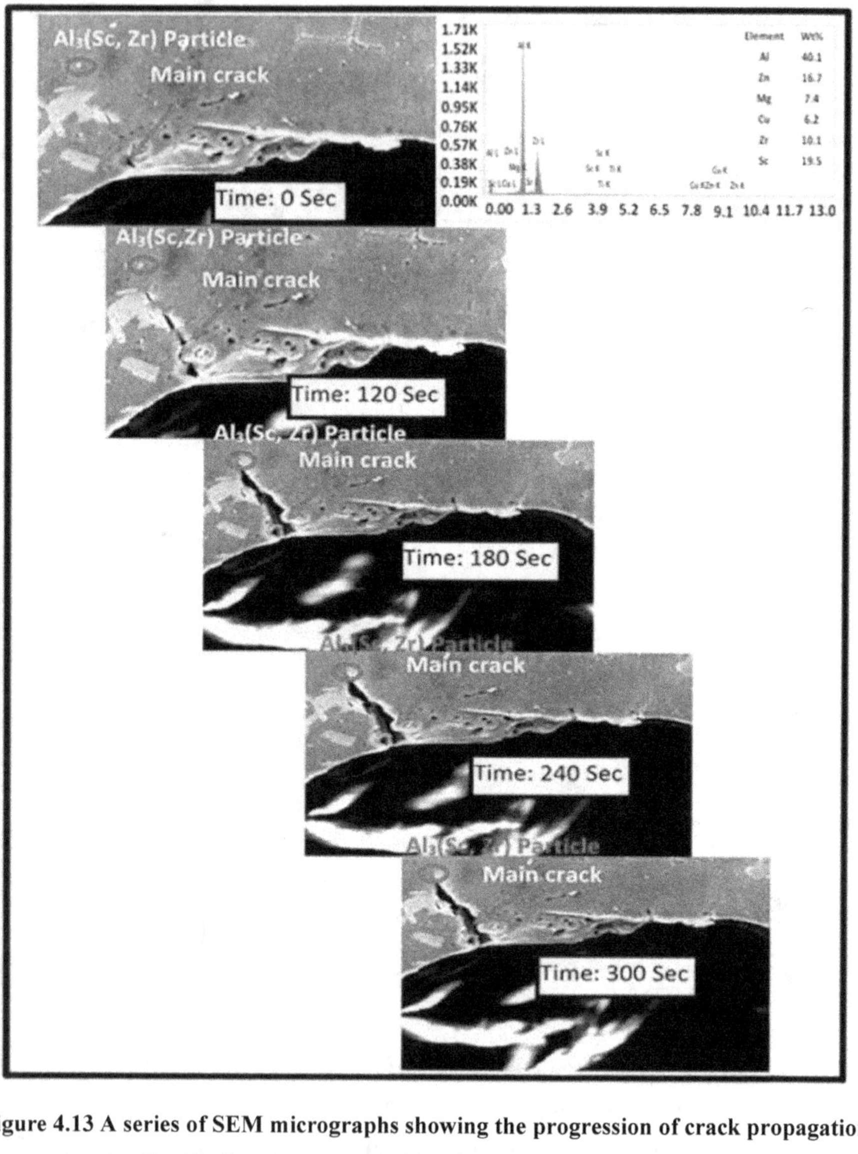

Figure 4.13 A series of SEM micrographs showing the progression of crack propagation over time in alloy D. Note the crack tip blunting occurring owing to the presence of Al_3(Sc, Zr) particle. Time from the start of crack propagation is also mentioned. EDX results confirm the presence of Al_3(Sc, Zr) phase.

Figure 4.10 shows how the 10-15 μm sized Al_3(Sc, Zr) primary particles pose considerable resistance to crack growth and cause crack tip blunting. This blunting phenomenon has been observed during the in-situ tensile tests (Figure 4.13), where the Al_3 (Sc, Zr) dispersoids considerably resist crack extension for almost 300 seconds after initiating cracks from the notch tip. Moreover, mechanisms such as crack meandering could also be seen in alloy D (Figure 4.12f). Collectively, these factors make alloy D significantly superior to all other alloys. The quasi-brittle behaviour (Figure 4.11) of the alloys under investigation is influenced by the size, shape, and distribution of primary intermetallic, documented in various earlier reports on dispersoid-reinforced alloys [24,44-45]. The crack propagation in these alloys shows a tendency towards micro-crack formation and their coalescence. A similar observation of microcrack coalescence in aluminium alloys was earlier observed by MishnaevskyJret al. [25]. However, the microcracks in the material are not only formed in the vicinity of the crack tip due to the stress field from the crack but also at sites throughout the stressed volume, consistent with the dislocation pile-up model for their nucleation [25,46]. The alloys A, B, and D microstructures showed the same behaviour in this respect, meaning energy considerations can be used to model the failure mechanism macroscopically. It is also observed that cast defects impact the initiation and propagation of cracks. In the second stage, areas with shrinkage porosity can be seen. These are mostly present in S and η′containing eutectics at the grain boundaries (Figure 4.7d). Similar observations have been reported by earlier authors [47-49]. The fracture surface also reveals porosity areas encompassing about 9% of its area, with pores acting as an artificial edge notch, causing stress concentration and crack initiation, followed by rapid quasi-brittle fracture [24]. The grain boundary S and η′ eutectics exhibit linear elastic failure behaviour, with microcracking occurring during tensile loading. The size and shape of the Al_3(Zr, Ti) dispersoids also impact the fracture strength of alloys B, C, and D, similar to other high-strength aluminum alloys [44, 50].

To further understand the process of cracks forming and propagating in front of a notch, a FEM study was conducted on a single-edge notch tensile test. According to the analysis, the stresses in front of the fracture point are in a certain direction relative to the notch, as shown in Figure 4.6. The stress generated reaches a peak value of 543MPa. The maximum equivalent stresses (Figure 4.6) are greater in the specified direction. These findings indicate that the local stresses are expected to surpass the externally applied stress and be roughly equivalent to the yield stress of the materials. Upon direct observation, it is evident that the microcracks are predominantly located in the direction of the highest equivalent stress (Figure 4.6 and 4.10). This indicates that there may have been a substantial accumulation of dislocations at the interface between the dispersoids and the matrix, resulting in the separation or detachment of the dispersoids from the matrix. Mishnaevsky Jr et al. [25] observed similar effects in Al-Si alloys. However, the alloys under investigation comprise coherent and incoherent phases governed by their dimensions, configuration, and dispersion. While the present work does not involve simulating coherency strains, the literature cited [51] emphasizes the crucial significance of understanding this phenomenon, particularly in the case of dispersoids distributed in a ductile matrix. The presence of small Al_3M type coherent precipitates in alloy D hinders the concentration of strain by effectively resisting high loads. This prevents the soft α-matrix from deforming unevenly. Conversely, several bulk incoherent particles lead to an escalation in localised strains, resulting in a higher probability of damage, as evidenced in alloy C. The incoherent particles, particularly the bulk Al_3(Zr, Ti) and grain boundary S-Al_2CuMg phases, are easily stressed when they align toward the highest stress. Intensely stressed areas can be seen at the extremities of bulk Al_3(Zr, Ti) phase or between the randomly distributed grain boundary particles (Figure 4.8 and 4.9). When these dispersoids are arranged so that their endpoints are close, an area of continuous deformation is generated. In the observed timeframe, it was found that the α-Al matrix region exhibits a higher strain compared to the surrounding eutectic colonies. This suggests that the eutectic colonies underwent less plastic deformation than the matrix regions, possibly attributed to variations in hardness between the matrix and eutectic colonies [24, 52]. The variation in strength is characterized by indentation testing of the different phases. A hardness of ~ 195 HV is found in the intermetallics and ~ 126 HV in eutectic regions in contrast to the α-Al matrix with a hardness of ~ 92 HV alloy D.

Figure 4.14 illustrates the schematic of various crack tip damage mechanisms seen during in-situ tests of the Al-6Zn-2.1Mg-2Cu-X alloy under monotonic tensile loading conditions. The nucleation of cracks takes place mainly from the notch tip (except in alloy C, where the crack is observed to initiate at the large agglomerated Ti and Zr-rich intermetallics) along the grain boundary; additionally, a few micro-cracks are found to have nucleated at randomly located positions (eutectic phases) in the vicinity of the notch tip. These micro-cracks join and form the main crack, which shows the mechanisms of crack growth as shown in the schematic (Figure 4.14). However, the in-situ tensile tests on the Al-6Zn-2.1Mg-2Cu-X alloy, especially when the microalloying conditions are changed, are therefore very important for understanding the deformation behaviour of the alloy. These tests are indispensable tools for unravelling the intricate mechanisms underlying crack tip damage in this Al-6Zn-2.1Mg-2Cu alloy with Sc, Zr and Ti additions (microalloying elements varying from 0.6 wt.% - 1 wt.%). Because alloy composition, microstructure, and mechanical properties affect each other in complex ways, in-situ tensile testing is the only way to fully understand and correlate them. By subjecting the alloy to controlled stress conditions in real-time, valuable insights into the material's behaviour at the microscale have been obtained, enabling a precise understanding of crack initiation, propagation, and eventual failure. This nuanced understanding is pivotal for optimizing material performance in diverse applications, from aerospace components to structural materials, ultimately fostering advances in materials design and engineering.

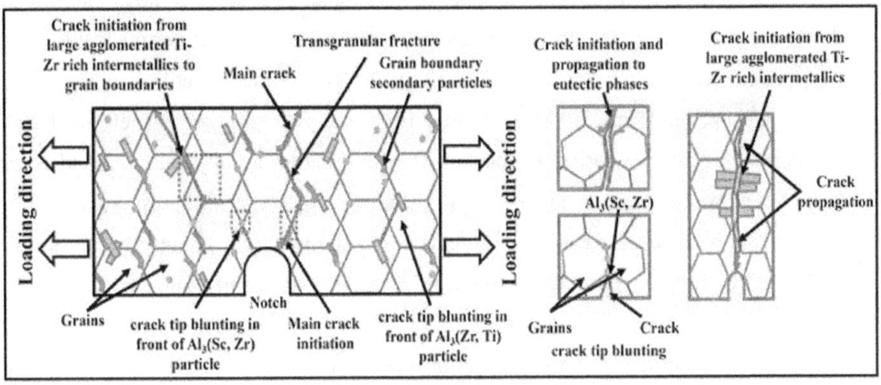

Figure 4.14 Schematic representations of different crack tip damage mechanisms prevalent in Al-Zn-Mg-Cu-X alloy.

4.5 Conclusions

The crack tip deformation behaviour of the Al-6Zn-2.1Mg-2Cu alloy with and without microalloying elements (Ti, Zr, and Sc) has been comprehensively explored for the first time through in-situ tensile testing under FESEM. This approach allows for a detailed interpretation of the intricate mechanisms involved in deformation and complex fracture behaviour. Grain boundary eutectic, intermetallic phases and dispersoids are primarily responsible for controlling the progression of crack initiation and growth in four different cast alloys. The following inferences can be drawn from this study:

1. Adding microalloying elements such as Zr, Ti, and Sc leads to a significant refinement in the grain structure. After optimising the Zr and Ti percentages, 0.25 wt.% Sc helps to reduce the grain size from 62 μm to 34 μm. This refinement is attributed to various intermetallics that form before solidification, including Al_3(Zr, Ti)and Al_3(Sc, Zr). However, the amount of alloying addition plays a vital role in the deformation behaviour of the developed alloys.
2. The limited ductility in alloy C is because the holding temperature during casting (i.e.; 745°C) is too low to dissolve the amount of 1 wt.% of Ti and Zr in the alloy. The particles grow in the melt and the coarseAl_3(Ti ,Zr) dispersoids fracture early. Hence,

it is important to keep the additions low enough that dispersoids do not grow in the melt during holding.

3. The micro-mechanism of fracture is inherently quasi-brittle in four different Al-6Zn-2.1Mg-2Cu-Xalloys. Strain accumulation in notched tensile specimens during in-situ testing determines the direction of fracture development behaviour. The tensile strength of these alloys is mainly determined by the cleavage fracture mechanism and particle/intermetallic decohesion. Additionally, the limited ductility in the matrix results from the formation of voids associated with second-phase particles and at the interface of coarse $Al_3(Zr, Ti)$ phases. Void nucleation usually starts around coarse particles at the grain boundary of S-Al_2CuMg and η-$MgZn_2$phases. Smaller grain boundary particles also cause crack deflection, making the crack path more tortuous.
4. The addition of the optimum amount of Ti, Zr, and Sc (named alloy D) leads to the formation of small coherent $Al_3(Sc, Zr)$, and $Al_3(Zr, Ti)$ dispersoids that help in arresting cracks and thereby increasing the crack resistance behaviour of alloy D. However, if the addition is too high as in alloy C, coarse $Al_3(Zr, Ti)$ intermetallics form and agglomerate which cause early fracture.

References

[1] T. Dursun, C. Soutis, Recent developments in advanced aircraft aluminium alloys, Mater. Des. 56 (2014)862-871. https://doi.org/10.1016/j.matdes.2013.12.002.

[2] X. Zhang, Y. Chen, J.Hu, Recent advances in the development of aerospace materials. Prog Aero Sci.97 (2018) 22-34. https://doi.org/10.1016/j.paerosci.2018.01.001.

[3] D. Dumont, A. Deschamps, Y. Brechet,A model for predicting fracture mode and toughness in7000 series aluminium alloys. Acta Mater. 52 (2004) 2529-2540.https://doi.org/10.1016/j.actamat.2004.01.044.

[4] Z. Huda, P. Edi, Materials selection in design of structures and engines of supersonic aircrafts: a review, Mater. Des. 46 (2013) 552-560. https://doi.org/10.1016/j.matdes.2012.10.001.

[5] R.G. Song, W. Dietzel, B.J. Zhang, W.J. Liu, M.K. Tseng, A. Atrens, Stress corrosion cracking and hydrogen embrittlement of an Al-Zn-Mg-Cu alloy, Acta Mater.52 (16) (2004) 4727-4743.https://doi.org/10.1016/j.actamat.2004.06.023.

[6] S.P. Knight, K. Pohl, N.J.H. Holroyd, N. Birbilis, P.A. Rometsch, B.C. Muddle, R. Goswami, S.P. Lynch, Some effects of alloy composition on stress corrosion cracking in Al-Zn-Mg-Cu alloys, Corros. Sci. 98 (2015) 50-62. https://doi.org/10.1016/j.corsci.2015.05.016.

[7] H.C. Fang, Luo, F.H. Luo, K.H. Chen, Effect of intermetallic phases and recrystallization on the corrosion and fracture behavior of an Al-Zn-Mg-Cu-Zr-Yb-Cr alloy. Mater. Sci. Eng. A.684 (2017) 480-490. https://doi.org/10.1016/j.msea.2016.12.009.

[8] S.H. Lee, J.G. Jung, S. Il Baik, S.H. Park, M.S. Kim, Y.K. Lee, K. Euh, Effects of Ti addition on the microstructure and mechanical properties of Al-Zn-Mg-Cu-Zr alloy, Mater. Sci. Eng. A. 801 (2021) 140437. https://doi.org/10.1016/j.msea.2020.140437.

[9] D. Mukherjee, M. Mukherjee, N. Mandal, S.K. Samanta, A. Maiti, Effect of micro-alloying on the characteristics of as-cast Al-Zn-Cu-Mg-X alloy with varying Cu and Zn, Mater. Today Commun. 34 (2023) 105445. https://doi.org/10.1016/j.mtcomm.2023.105445.

[10] M. Zhang, T. Liu, C. He, J. Ding, E. Liu, C. Shi, J. Li, N. Zhao, Evolution of microstructure and properties of Al-Zn-Mg-Cu-Sc-Zr alloy during aging treatment, J. Alloys Compd. 658 (2016) 946-951. https://doi.org/10.1016/j.jallcom.2015.10.296.

[11] Y. Wang, L.Cao, X. Wu, X.Lin, T.Yao, L. Peng, Multi-alloying effect of Ti, Mn, Cr, Zr, Er on the cast Al-Zn-Mg-Cu alloys, Mater. Charact. 201 (2023) 112984. https://doi.org/10.1016/j.matchar.2023.112984.

[12] J. Zhao, Z. Liu, S. Bai, D. Zeng, L. Luo, J. Wang, Effects of natural aging on the formation and strengthening effect of G.P. zones in a retrogression and re-aged Al-Zn-Mg-Cu alloy, J. Alloys Compd. 829 (2020) 154469. https://doi.org/10.1016/j.jallcom.2020.154469.

[13] P.Tan, J.Qin, X.Quan, D.Yi, B.Wang, Co-strengthening of the multi-phase precipitation in high-strength and toughness cast Al-Cu-Zn-Mg alloy via changing Zn/Mg ratios, Mater. Sci. Eng. A.873 (2023) 145024.https://doi.org/10.1016/j.msea.2023.145024.

[14] D. Bayoumy, W. Kan, X. Wu, Y. Zhu, A. Huang, The latest development of Sc-strengthened aluminum alloys by laser powder bed fusion, J. Mater. Sci. Technol. 149 (2023) 1-17. https://doi.org/10.1016/j.jmst.2022.11.028.

[15] D. Mukherjee, H. Roy, B. Chandrakanth, N. Mandal, S.K. Samanta, M. Mukherjee, Enhancing properties of Al-Zn-Mg-Cu alloy through microalloying and heat treatment, Mater. Chem. Phy., 314 (2024) 128881. https://doi.org/10.1016/j.matchemphys.2024.128881

[16] H. Jiang, H.Xing, Z. Xu, B. Yang, E.Liang, J. Zhang, B. Sun, Effect of Zn content and Sc, Zr addition on microstructure and mechanical properties of Al-Zn-Mg-Cu alloys. J.

Alloys Compd. 947 (2023) 169246. https://doi.org/10.1016/j.jallcom.2023.169246.

[17] J. Liu, P. Yao, N. Zhao, C. Shi, H. Li, X. Li, D. Xi, S. Yang, Effect of minor Sc and Zr on recrystallization behavior and mechanical properties of novel Al-Zn-Mg-Cu alloys, J. Alloys Compd. 657 (2016) 717-725. https://doi.org/10.1016/j.jallcom.2015.10.122.

[18] Z.Wang, Q.Pu, Y.Li, P. Xia, J. Geng, X. Li,M. Wang, D. Chen, H. Wang, Microstructures and mechanical properties of Al-Zn-Mg-Cu alloy with the combined addition of Ti and Zr, J. Mater. Res. Tech.22 (2023) 747-761.https://doi.org/10.1016/j.jmrt.2022.11.106.

[19] O. Wouters, J.T.M. De Hosson, Lead induced intergranular fracture in aluminum alloy AA6262, Mater. Sci. Eng. A. 361 (2003) 331-337. https://doi.org/10.1016/S0921-5093(03)00521-5.

[20] A.C. Reddy, S.S. Rajan, Influence of ageing, inclusions and voids on ductile fracture mechanism in commercial Al-alloys. Bull. Mater. Sci.28 (2005) 75-79. https://doi.org/10.1007/BF02711176.

[21] M. Vratnica, In-situ observation of the fracture process in Al-Zn-Mg-Cu alloys. Inter. J. Mater. Res.103 (2012) 624-632. https://doi.org/10.3139/146.110731.

[22] J.W. Li, J.P. Xie, W.Y. Wang, S.Z. Wei, Study on the in-situ SEM tension observation of E-A356 alloys, Mater. Sci. Forum. 561 (2007) 147-150. https://doi.org/10.4028/www.scientific.net/MSF.561-565.147.

[23] G.T. Motsi, P.A. Olubambi, T.J. Sono, L. Shoke, In situ electron microscopy studies on the tensile deformation mechanisms in aluminium 5083 alloy, Adv. Mater. Res. 1019 (2014) 103-111. https://doi.org/10.4028/www.scientific.net/AMR.1019.103.

[24] R.J. Lane, M.S. Kesler, K. Nawaz, R. Mirzaeifar, Investigating the failure behavior of cast Al-11Ce-0.4 Mg alloys using in-situ scanning electron microscopy tensile testing, J. Alloys Compd.947 (2023) 169491. https://doi.org/10.1016/j.jallcom.2023.169491.

[25] L.L. Mishnaevsky Jr, N. Lippmann, S. Schmauder, P. Gumbsch, In-situ observation of damage evolution and fracture in AlSi7Mg0.3 cast alloys. Engng. Fract. Mech.63 (1999) 395-411. https://doi.org/10.1016/S0013-7944(99)00027-2.

[26] J.W. Yeh, W.P. Liu, The cracking mechanism of silicon particles in an A357 aluminum alloy, Metall. Mater. Trans. A. 27 (1996) 3558-3568. https://doi.org/10.1007/BF02595447.

[27] D.L. McDowell, K. Gall, M.F. Horstemeyer, J. Fan, Microstructure-based fatigue modeling of cast A356-T6 alloy, Engng. Fract. Mech. 70 (2003) 49-80.

https://doi.org/10.1016/S0013-7944(02)00021-8.

[28] Z. Li, N. Limodin, A.Tandjaoui, P. Quaegebeur, J.F. Witz, D. Balloy, Influence of Fe content on the damage mechanism in A319 aluminum alloy: Tensile tests and digital image correlation. Eng. Frac. Mecha. 183 (2017) 94-108. https://doi.org/10.1016/j.engfracmech.2017.05.006.

[29] A. Bjurenstedt, E. Ghassemali, S. Seifeddine, A.K. Dahle, The effect of Fe-rich intermetallics on crack initiation in cast aluminium: An in-situ tensile study, Mater. Sci. Eng. A.756 (2019) 502-507.https://doi.org/10.1016/j.msea.2018.07.044.

[30] M. Besterci, O. Velgosová, J. Ivan, P. Hvizdoš, I. Kohútek, Influence of volume fraction on fracture mechanism of Al-Al_4C_3 system studied by "in-situ tensile test in SEM", Kovove Mater, 46 (2008) 139-143.

[31] B.Wisner, A. Kontsos, In situ monitoring of particle fracture in aluminium alloys, Fatigue & Fracture of Engineering Materials & Structures, 41 (2018) 581-596. https://doi.org/10.1111/ffe.12720.

[32] N.C. Adak, S. Chhetri, S. Sabarad, H. Roy, N.C.Murmu, P. Samanta, T. Kuila, Direct observation of micro delamination in graphene oxide incorporated carbon fiber/epoxy composite via in-situ tensile test, Compos. Sci. Technol. 177 (2019) 57-65. https://doi.org/10.1016/j.compscitech.2019.04.006.

[33] ANSYS Fluent Theory Guide. Release 19.0, R1 ANSYS, Inc., USA (2020).

[34] F.H. Chung, D.K. Smith, Industrial applications of x-ray diffraction, 1999. https://doi.org/10.1201/b16940.

[35] A.L. Ortiz, L. Shaw, X-ray diffraction analysis of a severely plastically deformed aluminum alloy, Acta Mater. 52 (2004) 2185–2197. https://doi.org/10.1016/j.actamat.2004.01.012.

[36] A. Rutecka, M. Kursa, K. Pietrzak, K. Kowalczyk-Gajewska, K. Makowska, M. Wyszkowski, Damage evolution in AA2124/SiC metal matrix composites under tension with consecutive unloadings, Arch. Civ. Mech. Eng. 20 (4) (2020) 135. https://doi.org/10.1007/s43452-020-00134-x.

[37] J.H. Chen, G.Z. Wang, R. Cao, J. Zhang, Study on notch fracture of TiAl alloys at room temperature. Metall. Mater. Trans. A, 35 (2004) 439-456. https://doi.org/10.1007/s11661-004-0355-6.

[38] D. Han, S. Cai, W. Sun, B. Yang, J. Wang, Strategies for Achieving Strength–Ductility Synergy in Face-Centered Cubic High-and Medium-Entropy Alloys, Adv. Eng. Mater.25 (2023) 2300291. https://doi.org/10.1002/adem.202300291

[39] X. Li, Q. Cai, B. Zhao, B. Liu, W. Li, Precipitation behaviors and properties of solution-aging Al-Zn-Mg-Cu alloy refined with TiN nanoparticles, J. Alloys Compd. 746 (2018) 462-470. https://doi.org/10.1016/j.jallcom.2018.02.271.

[40] T.S. Liu, F. Qiu, B.X. Dong, R. Geng, M. Zha, H.Y. Yang, S.L. Shu, Q.C. Jiang, Role of trace nanoparticles in establishing fully optimized microstructure configuration of cold-rolled Al alloy, Mater. Des. 206 (2021) 109743. https://doi.org/10.1016/j.matdes.2021.109743.

[41] M.V. Karpets, Y.V. Milma, O.M. Barabash, N. P. Korzhova, O.N. Senkov, D.B. Miracle, T.N. Legkaya, I.V. Voskoboynik, The influence of Zr alloying on the structure and properties of Al_3Ti, Intermetallics. 11(2003) 241-249. https://doi.org/10.1016/S0966-9795(02)00234-0.

[42] X.G. Chen, M. Fortier, TiAlSi intermetallic formation and its impact on the casting processing in Al-Si alloys. J. Mater. Process. Technol. 210(13) (2010) 1780-1786. https://doi.org/10.1016/j.jmatprotec.2010.06.009.

[43] J.R. Castillo-Sánchez, G. Salloum-Abou-Jaoude, A.E. Gheribi, P. Lafaye, K. Oishi, J.P.Masse, E. Bousser, G. L'Espérance, J.P. Harvey, Synthesis and characterization of $(Al, Si)_3(Zr, Ti)$-$D0_{22}/D0_{23}$ intermetallics: Understanding the stability of silicon substitution. Acta Materialia, 262 (2024) 119455. https://doi.org/10.1016/j.actamat.2023.119455.

[44] S. Papanikolaou, P. Shanthraj, J. Thibault, C. Woodward, F. Roters, Brittle to quasi-brittle transition and crack initiation precursors in crystals with structural Inhomogeneities, Mater. Theory 3 (2019) 5. https://doi.org/10.1186/s41313-019-0017-0.

[45] J.M. Byun, S.-R. Bang, C.W. Park, M.-J. Suk, Y.D. Kim, Structural size effects of intermetallic compounds on the mechanical properties of Mo-Si-B alloy: an experimental investigation, Met. Mater. Int. 22 (2016) 81-86. https://doi.org/10.1007/s12540-015-5497-0.

[46] V.V. Panasyuk, A.E. Andreikiv, V.S. Kharin, Origin and growth of microcracks produced by blocked accumulations of dislocations. Mater Sci. 21 (1985) 105-115. https://doi.org/10.1007/BF01150625.

[47] K. D, L. Kovačević, P. Terek, D. Krumes, Z. Kolumbić, Use of SEM and EDS analysis in the investigation of Al-Si-Cu piston alloy cast Porosity, Metalurgija 48 (2009). https://hrcak.srce.hr/35472

[48] A. Samuel, Y. Zedan, H. Doty, V. Songmene, F.H. Samuel, A review study on the main

sources of porosity in Al-Si cast alloys, Adv. Mater. Sci. Eng. 2021 (2021) 1921603. https://doi.org/10.1155/2021/1921603

[49] J. Campbell, J. Campbell (Ed.), Chapter 6 - Gas porosity, in Castings, Second ed., Butterworth-Heinemann, Oxford, 2003, pp. 178-204.

[50] A. Raina, Size-Dependent fracture characteristics of intermetallic alloys, Exp. Mech. 62 (5) (2022) 863-877. https://doi.org/10.1007/s11340-022-00831-z.

[51] M. Zain-ul-abdein, D. Nélias, Effect of coherent and incoherent precipitates upon the stress and strain fields of 6xxx aluminium alloys: a numerical analysis. Int J Mech Mater Des. 12 (2016) 255-271. https://doi.org/10.1007/s10999-015-9298-x.

[52] H. Kaya, U. Böyük, E. Çadırlı, N. Maraşlı, Measurements of the microhardness, electrical and thermal properties of the Al–Ni eutectic alloy, Mater. Des. 34 (2012) 707–712. https://doi.org/10.1016/j.matdes.2011.05.030.

www.ingramcontent.com/pod-product-compliance
Lightning Source LLC
LaVergne TN
LVHW010725061126
851857LV00014B/1344
* 9 7 8 1 9 6 2 1 1 6 7 8 7 *